Astronomers' Observing Guides

WITHDRAWN

Steven R. Coe

Nebulae
and How to
Observe Them

 Springer

Steven Coe
Phoenix, AZ 85022
Stevecoe@ngcic.org

Cover illustration: The Sword of Orion with Takahashi refractor. Courtesy of Jon Christensen.

Library of Congress Control Number: 2006926453

ISBN-10: 1-84628-482-1
ISBN-13: 978-1-84628-482-3

Printed on acid-free paper.

© 2007 Springer Science+Business Media, LLC

9 8 7 6 5 4 3 2 1

springer.com

Acknowledgments

This book is dedicated to the two groups of people who have supported me without hesitation in observing the sky and writing about my adventures.

First is my family. Fred Rainey, my grandfather, was the first to show me the sky as I learned a few stars and constellations when we were out fishing in the early morning hours, during my childhood. Linda Ross, my wife, has provided me the time to enjoy the Universe and the quiet I needed to enter it into the computer. Without her love and devotion I could not have completed one book, much less two. Her sister Laura Ross and my brother-in-law Bob Lambert provided me some time at their cabin so that I could put it all together and create a coherent manuscript out of the mounds of notes that I had written at the eyepiece. Much of the rest of my extended family: Audrey, Judy and Matt, Meghan, McKenzie, Ashley and Kevin all asked how the book was going and were willing to listen as I went on about my progress. Well, it is done!

The second group is the Saguaro Astronomy Club members who have been my observing buddies for many years. A.J. Crayon, David Fredericksen, Curt Taylor, Rich Walker, Tom and Jennifer Polakis, Pierre Schwaar, Bob Erdmann, George deLange, Chris and Dawn Schur, Matt Luttinen, Thad Robosson, Rick Rotramel, Paul Lind and many others have been out in the dark with me, for decades in many cases. They have been generous enough to allow me to direct the observing for a while when I needed observations. We have shared victories at finding a faint object that none of us had seen before and defeats when giving up the chase after half an hour of observing blank sky. I believe that it made us stronger and cemented our friendship.

I appreciate all of you
Steve Coe

Contents

Contents

Contents

Chapter 1

Viewing the Sky

This book is about nebulae, those clouds of gas and dust that are the beginnings and endings in the lifetime of stars. We will get into much more detail about that later. Right now, I would like to introduce myself and discuss the reasons for this book and some rudimentary information about the equipment you will need to observe the Universe.

I really don't remember a time when I did not know at least a few of the brighter stars and constellations. While going fishing with my grandfather early in the morning, he would point out a few bright lights in the night sky and see if I remembered them the next time we were fishing. This simple task started me down the path of knowledge about the nighttime light show that is always overhead.

After a hitch in the US Navy as a submarine sailor, I decided to become a professional astronomer in 1976. So I enrolled at Arizona State University and quickly found that I had made a mistake. I thought that I was going to get a chance to *look* through those giant telescopes; that virtually never happens with modern telescopes. I also found that most of my astronomy professors were mathematicians first and astronomers second.

Fortunately, I found a job that paid well and had time off to go out with the telescope. For 25 years I taught electronics at DeVry Institute of Technology in Phoenix, Arizona, USA. I am also a long-time member of the Saguaro Astronomy Club (SAC), one of the most active groups of observers in the world. With those two bits of good fortune in place, I started observing the sky with a wide variety of telescopes.

I have had lots of fun observing both on my own and with the members of SAC over the past 25 years or so. Getting out under clear desert skies and enjoying the view has been a source of joy for me. In this book, I plan to share that with you.

Another reason for writing this book is to give telescope owners some ideas about how to get better at observing. There are lots of telescopes just sitting in a closet because their owners didn't know how to improve their observing skills. If you just glance at several nebulae, they start to look the same. Once you learn to really observe, then you will see that many are unique. Getting better at seeing interesting detail in the field of view of your telescope is worth doing. It just takes some time with your telescope and some effort to improve your observing skill. This book will help.

Binoculars

I doubt that the first device you thought of for viewing the night sky was a pair of binoculars. However, they have some advantages.

First is ease of use. A pair of binoculars is easy to carry outside and once you have learned to get them in focus, they are easy to use. If you have rarely used binoculars before, focus them on a distant object during the day and then they will be close to your point of focus that evening. Most modern binoculars have one focus wheel in the center that moves both eyepieces, and then one eyepiece has an independent focus. I close one eye and then the other as I use the two focus wheels. Once you are in focus, both eyes will show a sharply focused scene; then open both eyes for the binocular effect.

Second is the wide field of view that binoculars will provide compared to a telescope. Some of the brightest objects in the night sky are also large in size. So the binoculars allow you to frame that nebula or star cluster with some space around it, a most pleasing view.

When we go out observing I always take my binoculars and use them for a break. When I am ready to get off my feet for a few minutes, I pull out the binoculars, sit in the comfortable folding chair and scan the sky.

The numbers associated with a pair of binoculars are telling you two important facts about them. For instance a pair of binoculars may be 10X50 or 7X35; this is spoken as "ten by fifty or seven by thirty five." The first number is the amount of magnification and the second is the size of the objective glass in millimeters. A small pair of binoculars, such as the two sizes mentioned above, is a great place to start. These types of binoculars are easy to use both day and night and it is easy to find things in the night sky using these modest binoculars. There are several books on binocular observing available if you wish to learn more.

Telescopes

I know that you want to race out right now and buy a telescope, but please wait a few minutes, at least. A little thought will let you make a good decision. Telescopes are generally spoken of by the aperture of the lens or mirror that is gathering light. When someone asks about your scope your might say: "It's a 10 inch Newtonian" or "This is a 120 millimeter refractor." Another piece of information about a telescope is its focal ratio. This is the ratio of the size of the aperture to the length of the tube. So an f/6 telescope is six times longer than it is wide. Generally, telescopes with a small value of f/ratio do a good job of providing a wide field of view, and longer scopes work well at high magnification.

The two main types of telescopes are refractors and reflectors. These names are similar and so you need to look carefully.

Refractors are telescopes that use a lens to bend, or refract, the light toward a point of focus. The good news about refractors is that they provide a sharp field of view that provides lots of contrast between the light and dark areas in the view. This is what we are looking for when trying to chase down a faint nebula that is glowing among the stars in the field. The bad news is that a good refractor is expensive for the size of the telescope. Why do you think that there are lots of 4 inch (100 mm) refractors on the market? Because the cost of a well-made 6 inch (150 mm) or larger refractor is very high indeed.

Me and a 120 mm f/8 refractor.

Reflectors use a mirror to reflect the light to a viewing location. The Newtonian layout brings the focal point out the side of the tube, while the Cassegrain variants bring the focal point through a hole in the mirror and then out the back of the telescope.

I have built and owned several Newtonians over the years and they have served me well. Newtonians have the largest aperture you can buy for the money and that is always good. The bad news is that they demand some skill at keeping the mirrors aligned and clean.

My old 10 inch (250 mm) Newtonian reflector. C'mon how many people have a photo of a bride looking through their telescope? I just had to use it.

My large telescope today is a Schmidt-Cassegrain. This telescope is generally a Cassegrain system, but incorporates a corrector plate at the front to sharpen the image. The good news is that these scopes provide a lot of aperture and focal length in a small package. The bad news is that they are more expensive than a Newtonian of equal aperture.

Some of the observations in this book were made with a telescope that is a combination of the two types, the Maksutov-Newtonian. The "Mak-Newt" is generally a Newtonian layout, with a corrector lens at the front. I purchased a 6 inch f/6 version of this type of telescope from the estate of my friend Curt Taylor. He had bought it with the idea of using it to view the planets at high power, and it did a fine job of that. I also discovered that it did a terrific job of providing excellent wide-field views of the Milky Way. It is the only telescope that I have ever owned that I wish I had not sold. But I will have another someday – don't tell my wife I said that.

Right now I have two telescopes that really do show the sky very nicely. My big scope is a Celestron Nexstar 11, a Schmidt-Cassegrain telescope (SCT) that provides 11 inches of aperture or 280 millimeters to most of the world. I enjoy the ease of use of the GOTO electronics system, but it also provides excellent viewing. I am impressed with my NS 11.

The only thing that a big SCT will not provide is a wide field of view. So I traded some equipment at Riverside Telescope Makers' Conference and wound up with a

My 6 inch (150 mm) Maksutov-Newtonian.

Me and the Nexstar 11 inch.

4 inch (100 mm) f/6 refractor from Orion Telescopes. I have also been impressed with this telescope. It provides a pretty sharp field of view and a wide field. Andrew Cooper in Tucson was nice enough to make me a piggyback plate that fits on the NS 11 and lets the rich field refractor mount on the back. I can also put the refractor on a driven German equatorial mount and take photographs.

The 4 inch (100 mm) f/6 refractor; dark skies near the little town of Happy Jack, Arizona.

Those two telescopes provide a very nice tandem set of instruments and give me both wide field, low magnification in the small refractor or a high-power view in the big scope.

I mentioned the 6 inch f/6 Mak-Newt; I also have owned a 13 inch f/5.6 Newtonian and a 10 inch f/5 Newtonian. So over the past 20 years I have owned a wide variety of telescopes. I have also been lucky enough to have several friends with large instruments who have allowed me to use their telescopes. With all that, I have amassed observations with a wide variety of telescopes. I certainly plan to share those with you.

Eyepieces

Well, now that you are the proud owner of that new telescope, you will find the truth about being an "amateur" astronomer. The truth is: you can always spend money on accessories. Just like every sports car needs expensive fuel, every telescope needs eyepieces.

So let's take a look at the wide variety of eyepieces that are available to modern amateur astronomers. If you get a group of observers together and let them chat for long enough, the subject will become eyepieces sooner or later. The reason is that eyepieces are the most important accessory that you can purchase for your telescope. They are also the most personal. You just have to try them out and see if your eye matches the optics within the eyepiece. Sometimes it will and sometimes it won't. A.J. Crayon has been my observing buddy for over 25 years and he wisely says "Eyepieces are a religious discussion." I agree completely.

Once the scope itself has been purchased, then all the characteristics of that astronomical viewing system are determined by the eyepiece. I always look at an eyepiece as a small magnifying microscope which allows me to inspect the image formed by the mirror in my telescope. If you aim your scope at the Moon, then your optical system will create a tiny image of the Moon in midair; your eyepieces let you observe that image with varying magnifications and fields of view, all by changing the eyepiece.

Changing the focal length of your eyepiece changes the magnification and field of view.

There are lots of choices when it comes to eyepieces. Notice that some of the eyepieces on the right are 1.25 inch, one is 2 inch only and one will insert into either focuser size.

Let us begin by getting some terminology straight. Here is a list of the definitions of some words used in connection with eyepieces:

- Apparent field of view – this is the width in degrees of the field as seen through just the eyepiece alone. If I have two eyepieces with the same focal length, the one with the larger apparent field of view will show more of the sky if inserted into the same telescope. This parameter is determined by the design of the lenses inside an eyepiece.
- Curvature of field – good eyepieces provide a field of view which is flat. The focused image should be sharp from edge to edge. Star fields are a tough test of this characteristic.
- Distortion – good eyepieces also have little distortion. This means if you viewed a piece of lined graph paper that all the lines would be straight and would cross at right angles. Distortion can be a problem for only a small section of the field of view, but curvature generally happens to the entire field of an eyepiece.
- Exit pupil – the lenses in an eyepiece form an image that floats in midair just outside the lens closest to your eye. When you observe you place your eye so that it can see this exit pupil image. If all is going as planned, the image size will fit with room to spare within your eye. The size of this image is the exit pupil.
- Eye relief – the distance from the eye lens to your eyeball. This value is important to eyeglass wearers. If you need to have your glasses on to view the sky, there must be plenty of eye relief so that your eyeglasses will fit between the eyepiece and your face. Those of us who don't wear glasses to observe will appreciate some eye relief to avoid the feeling that we are jamming our eye lens against the glass lens of the eyepiece.
- Focal length – the apparent distance from the lens to the object being viewed, in this case the image formed by your telescope. Long focal length eyepieces show a large portion of the image being viewed and short focal length eyepieces will allow a small section of the image to be inspected. This is how you choose the magnification of your optical system. Pick out a long focal length eyepiece, say 40 mm to 24 mm, and the system will give a wide field and low power. Select a short focal length eyepiece, around 8 mm to 4 mm, and you will get a high-power, small field of view look at whatever is in the scope.

- Ghost images – in poorly made eyepieces some of the light from a bright star can reflect about within an eyepiece and form faint images within the field of view. These ghost images can be subdued by multicoating the lenses in the eyepieces. Only the cheapest eyepieces nowadays are not coated to suppress this problem.
- True field of view – this is the field of view of the entire telescope system, including the eyepiece.

Now that we know the meanings of some key phrases, let us move gently into a little calculation concerning eyepieces. There are three formulae that apply to using and understanding the values associated with eyepieces. These formulae are:

Magnification = Telescope focal length/Eyepiece focal length
Exit pupil = Telescope aperture/Magnification
True field of view = Apparent field of view/Magnification

Just remember that one inch equals 2.5 cm or 25 mm and you are ready to figure out these values for your telescope. So grab your calculator and we will try a worked example of some scope and eyepiece combinations. Assume you have a 6 inch f/8 telescope. That means the scope has 48 inches of focal length from 6 inches times f/8. Converting 48 inches to millimeters equals 1200 mm of focal length from 48 inches X25 mm/in.

Let's say you have three eyepieces which have focal lengths of 20 mm, 12 mm and 7 mm. Here are the magnifications each will supply:

60X for the 20 mm eyepiece from 1200 mm/20 mm
100X for the 12 mm eyepiece from 1200 mm/12 mm
171X for the 7 mm eyepiece from 1200 mm/7 mm

Now, here are the exit pupils for those eyepieces. Remember, you had to convert 6 inches to 150 mm first.

2.5 mm exit pupil for the 20 mm eyepiece from 150 mm/60X
1.5 mm exit pupil for the 12 mm eyepiece from 150 mm/100X
0.88 mm exit pupil for the 7 mm eyepiece from 150 mm/171X

To figure out the true field of view (FOV) for each eyepiece, we need to know the apparent field of view for the type of eyepiece used. Let's assume you are evaluating eyepieces with an apparent field of 60 degrees.

1 degree FOV for the 20 mm eyepiece from 60 degrees/60X
0.6 degree FOV for the 12 mm eyepiece from 60 degrees/100X
0.35 degree FOV for the 7 mm eyepiece from 60 degrees/171X

Because the true FOV is often less than 1 degree, this value is generally given in arc minutes. There are 60 arc minutes in 1 degree. So, 0.6 degrees X60 arc minutes per degree equals 36 arc minutes as the true FOV of the 12 mm eyepiece. Also, 0.35X60 means that the 7 mm eyepiece provides a 21 arc minute field.

I know that all this math is not particularly fun, but it does give some useful results. We can draw some general conclusions from our results. As the power is increased in your telescope, you get a smaller exit pupil and a narrower field of view. Because the pupil of your eye cannot get wider than 7 mm, it is not useful to buy an eyepiece that gives a larger exit pupil than that. The news is worse for those of us in advanced puberty. If you are over 35 years old, your eye probably does not open larger than 6 mm. At the other end of the scale, magnifications which yield an exit pupil smaller than 0.5 mm are not very useful either. It turns out that your eye has its best resolution if provided with an exit pupil of about 2 mm. So every set of eyepieces should provide a magnification that gets the system close to this value.

So if you are looking for great, wide-angle views of Eta Carinae or the Orion Nebula, then get an eyepiece with a long focal length and a wide apparent field of view. Are you looking for Encke's division in Saturn's rings? Or maybe trying to spot detail within planetary nebulae? Then look for an eyepiece with a short focal length and still with good eye relief so you don't have to jam your eye into the glass to observe. Maybe you wish to observe star clusters and galaxies at their best: then get an eyepiece that provides that magic 2 mm exit pupil.

In the same way that there are several different designs of telescope optics, the lenses within an eyepiece are arranged in a variety of ways to provide the observer with magnification and a focused field of view to observe. This section will look into the types of eyepieces available to modern observers and the strengths and weaknesses of each design.

Basically, as lens makers got better at making consistent glass with the same curves each time, they realized that adding more lenses generally reduced the aberrations in the eyepiece. Many of the designs I will discuss are named for the person who invented the combinations of lenses which make up these eyepieces.

So the first eyepiece designs, the Ramsden and Huygenian, only contain two lenses and are very poor eyepieces by modern standards. They have very narrow fields of view, short eye relief and many aberrations. Cheap telescopes often include these inexpensive eyepieces.

The Kellner is the best of the inexpensive eyepieces. This style of lens has been around for many years and it contains one doublet (two lenses together) and one singlet lens for a total of three pieces of glass inside. The Kellner does not have any excellent characteristics, but it also has few real flaws. Kellner eyepieces have decent eye relief, a modest field of view (45 degrees) and little curvature of field.

The Plossl eyepiece is composed of two doublets, which are identical to each other. For this reason, you will also hear it called a symmetrical eyepiece. It is a good eyepiece and many observers look no farther than a good set of Plossls. They have a medium field of view (55 degrees), good eye relief and are well corrected for aberrations. They cost more than Kellners, but they are worth it.

Orthoscopic eyepieces are generally not named for their inventors, Mittenzwey and Abbe, and I think you can see why. The "Orthos" have one outstanding characteristic: the aberrations and distortions in these eyepieces are virtually non-existent. These flat-field eyepieces have mediocre eye relief and field of view (45 degrees). This design contains a triplet lens with one singlet nearest your eye.

The Erfle eyepiece was invented to provide a wide apparent field of view and they do that (65 degrees). What the Erfle design gives up is some sharpness of the image at the edge of the field of view. Also, if there is a very bright star nearby where you are observing, some ghost images can appear within that wide field. Inside the Erfle is a combination of three doublet lenses.

This is where the eyepiece world stood for many years. Then the advent of computerized lens designs changed the standards for eyepiece manufacturers.

Enter the Nagler and Ultra Wide designs. These computer-designed eyepieces contain either seven or eight lenses, some with curves ground into them which would have been impossible before modern glass-polishing machines were constructed. These designs provide an extremely wide field of view (82 degrees) and low distortion fields at those wide angles. They all have two disadvantages: cost and weight. All that glass is going to cost more to grind and put together. Also, once they are assembled, these eyepieces weigh nearly 2 pounds in long focal lengths.

Along with eyepieces themselves, there is a device which will change the magnification of your system. It is called the Barlow lens. Just slide your eyepiece into the Barlow and put the whole thing into the eyepiece focuser and you have raised the magnification. The good news is that the eye relief of the system is the eye relief of the eyepiece alone. So for high power it is much easier to use a 10 mm eyepiece and a 2X Barlow than it is to use a 5 mm eyepiece and its short eye relief.

This is a great idea and I have owned a Barlow since my first scope, but there are limits. I find that Barlows which more than double the power are also introducing too many optical aberrations to the viewing system to allow me to believe I am seeing more detail than I saw without the Barlow in place. So use your Barlow in moderation and purchase a Barlow that magnifies either 1.8X or 2X and it will prove a very useful device.

Now that you have acquired all this knowledge about eyepieces, I'll bet you are still left with the same question: "Which ones do I buy?" That is a tough query, but I will give you my opinion. If you are just getting started, purchase three eyepieces. Buy one low-power, wide-field eyepiece which has a focal length between 35 mm and 25 mm. Get one medium-power eyepiece, from 20 mm to 12 mm. Buy one high-power eyepiece, from 9 mm to 6 mm focal length. Later, you can either get a Barlow lens with a magnification of 1.8X or 2X or "fill in" the focal lengths you have chosen. With one each of low-power, medium-power and high-power eyepieces, you are prepared to observe a wide variety of what there is to see in the sky.

As time goes by you can fill in as much as your budget will allow. You might choose a really wide-field 40 mm eyepiece, or maybe something between the medium and the high-power. I know that if you are just getting started, you might be thinking about a very high-power eyepiece in the range of 4 mm focal length. Even though it seems nifty to have a scope that can go to 600X, the number of evenings steady enough to use extreme magnifications is rare. You can make use of very high power occasionally, but not often.

What design of eyepiece to purchase is the subject of much talk when astronomers compare eyepieces and determine how much money they wish to spend. If you can afford it, at least start with a medium power Plossl, a high-power Orthoscopic, and a wide-field Erfle. If you are really in a pinch for money then Kellners will suffice. However, if you go out and observe with other folks who have better eyepieces than yours, it can be an expensive trip. One spectacular view through someone's brand-new pride and joy eyepiece can have you looking through catalogs and checking the limit on your credit card.

I have had a variety of different eyepieces in the 20 years that I have been using telescopes. My first scope was an 8 inch f/6 and it had a 1.25 inch focuser, so all my eyepieces were that size. I used three Erfles: 20 mm, 16 mm, and 12 mm for medium-power viewing. When I first got the scope, I did what I have told you not to do; I ordered it with a 4 mm eyepiece and never saw a clear view in it. I was able to trade the 4 mm for a 6 mm Orthoscopic that became a prized eyepiece for looking at fine detail on the Moon and the planets. Once I added a 2X Barlow, I was set, and my eyepiece collection changed little for several years.

When I sold the 8 inch to finance a 17.5 inch Dobsonian (yes, aperture fever got me too), I decided I needed a 2 inch focuser and an eyepiece to fit. Luckily, I found a war surplus 38 mm Erfle that only needed some machining to make a sleeve that fit the 2 inch focuser. A friend with a lathe made the part and I was in business. Again, my eyepiece collection seemed complete for a while.

It is 1981, me and my 17.5 inch (450 mm) Newtonian. This was my first trip to really dark skies.

In the 80s the Nagler revolution hit. The first 13 mm Nagler eyepieces I used had a problem that was serious for some observers, including me. Some folks see a "kidney bean," a dark marking within the field of view, which will not go away regardless how the observer moves their eye or head. I did not view this as a problem, because it prevented me from spending the money for these expensive eyepieces.

However, Meade decided to release its Ultra Wide series and I got a chance to use the 14 mm at the Riverside Telescope Makers' Conference. That was the final straw. The wide field of view, generous eye relief and excellent contrast of these eyepieces sold me. I found someone to purchase my old eyepieces and I completed the set of Ultra Wide eyepieces. In the same time, I also used and then bought a 22 mm Panoptic eyepiece that was excellent in my 13 inch f/5.6 Newtonian. The wide, flat, contrasty field of view of the Panoptic is stunning and it has become one of my favorite eyepieces.

Once I had the 22 mm Panoptic, I had a chance to observe with the 35 mm Panoptic (ain't love grand?). As I mentioned in my last book (*Deep Sky Observing – The Astronomical Tourist*, also published by Springer), I planned to use the royalties from that book to buy a 35 mm Panoptic and I have done that. I thank all of you who purchased that book and contributed to my eyepiece collection.

I have made one other change to my eyepiece assortment over the years. I have stopped using Barlow lenses. High-power eyepieces have been designed that provide lots of eye relief and are now much more comfortable to use compared to older designs. I have several of the Lanthanum eyepieces; they provide about 20 mm of eye relief and a sharp and contrasty field of view. Because I can now purchase a 5 mm eyepiece that gives high power and does not feel like I am trying to jam my eye into the glass, I have no need of a Barlow.

One of the good things about traveling to an astronomy conference is the opportunity to see and use a variety of equipment. One of my observing friends, David

Fredericksen, and I have been going to the Riverside Telescope Makers' Conference for almost three decades. We have had a chance to meet and chat with a wide variety of observers and see what they are doing. Three years ago (it is 2006 as I write this) we talked with one of the Celestron engineers about how the Nexstar 11 is manufactured and then we had a chance to use it under the stars. Consequently, David and I have both purchased a Nexstar 11 GPS and are very satisfied with the optical and mechanical performance. Every big astronomy meeting has a swap meet associated with it and you can find some real bargains on a wide variety of telescopes and accessories. I highly recommend making the time to attend a big "star party."

Make certain that your eyepieces are well cared for. Get some foam padding and cut out cavities to fit the eyepieces and protect them. Be careful when cleaning your eyepieces. Never rub them with any real force, always gently, or you will scratch the coating. Use a squeeze bulb or canned air to blow off any dust before cleaning. I use a special cleaning cloth I purchased at a camera store which does an excellent job removing the greasy fingerprints which inevitably happen. If you are thoughtful in purchasing good eyepieces and then protecting them from the elements, they will last many years and provide you with spectacular views of the heavens.

What's to See – Stars, Galaxies, Nebulae

OK, you are the owner of that new telescope and you have done you homework on eyepieces and are happy with the equipment you have purchased. Now the question is: "What can I see with all this stuff?."

All of the observing targets in this book are about objects beyond the Solar System, so they are called "deep sky" objects. The most obvious deep sky objects to observe are stars. Stars are glowing balls of gas that shine because the core of the star is so hot and dense that there is a fusion reaction going on in the middle of the star. The fusion reaction takes low-mass atoms (elements like hydrogen and helium) and fuses them together to create heavier elements (oxygen, calcium, and iron). Our Sun is a rather ordinary star.

To me, the most amazing fact about our Universe is that all the oxygen you are breathing, all the calcium in your teeth and all the iron in your blood were manufactured within stars. The second most amazing fact is that humanity figured that out.

As we learned more about the size, temperature and brightness of stars they were given a designation using the letters of the alphabet. In order to talk about something, we first need to invent a language. Because we knew so little about stars in the nineteenth century, it turns out that the temperature and size sequence of stars is not the same as the alphabet sequence. So, in order of temperature from hottest to coolest, the sequence is: O, B, A, F, G, K, M. Our Sun is a rather plain G2 star. Fortunately, they last a long time and don't change their energy output much over that lifetime. The hot O and B stars are going to be important in discussing nebulae and how they shine.

Just putting in a wide-field eyepiece and scanning around the sky you will see that many stars are grouped together in star clusters. The gravity of these stars pulls them together and holds them in the cluster. This book will cover several nebulae that surround a star cluster.

The Pleiades, a star cluster in the constellation of Taurus. This cluster includes a glowing nebula among the stars. Photo by Chris Schur with an 8 inch Schmidt camera.

The nebulae are clouds of dust and gas that exist between the stars. They appear as a fuzzy cloud in the eyepiece of the telescope. Stars have a life cycle: they are born, live, and die. Nebulae play a critical role in that life cycle. The place where stars first light off and begin to shine is within gaseous nebulae. As stars get toward the end of their life, they give up their dust and gas back into space to be used again. The chapter on "Nebula Knowledge" will cover this in more detail.

Galaxies are the huge "star cities" that gather together stars and nebulae into massive conglomerations. Again, the gravity of the stars is what keeps the galaxy together. When you are far from the city lights, you can see that glow across the sky that is called the Milky Way, which is the galaxy in which we live. The glow of the Milky Way is the combined light of millions of stars. Galaxies are the most obvious way in which the Universe organizes itself. Just as there are star clusters, there are galaxy clusters as well. Remember, gravity is pulling everything together. The only place that "anti-gravity" exists is on *Star Trek*, sorry.

There will be other books in this series that will cover a variety of astronomical objects. This book is going to concentrate on nebulae, what they are, and what can be seen at the eyepiece of a telescope.

Here is a shot of the famous Whirlpool Galaxy by Chris Schur with a 12.5 inch f/5 Newtonian.

Chapter 2

Becoming a Deep Sky Observer

Beyond the obvious of being able to set up your telescope and get it ready for observing, there are some other things you will need to do so you can observe the sky at its best. This chapter will cover some things to think about as you consider how to observe the deep sky.

Finding a Site

Since most of humanity now live in cities of some size, you will need to find an observing site that will get you away from the lights, smog and traffic of the city. All of the best astronomy sites are far from the nearest city.

My observing friends and I have generally split our observing sites into one of two types. "Close in" sites are about 40 to 60 miles (60 to 90 km) from the city. That way, we can leave in the afternoon, arrive at the site before sunset, set up the telescopes, observe for several hours and then dismantle the telescopes, pack up and drive back home around midnight or 1 AM. Our "distant" sites are about twice the distance from the city lights and generally demand that we stay overnight. If we are traveling to a distant site, then we try and make it a two-night trip to maximize our observing time.

The good news about the United States and Canada is that there are parks of a variety of types and also much public land in the western states to provide places to set up the telescope and enjoy the night sky. I understand that in the eastern

The Saguaro Astronomy Club set up at Five Mile Meadow, near the town of Happy Jack, Arizona. It is very far from the lights of Phoenix and the viewing is excellent at this high altitude of 6800 feet (2100 meters).

United States and much of Europe that is not the case. So many people may have limited options when it comes to choosing a site. This is where a local astronomy club comes in. Many clubs are built around an observatory that is maintained and used by the members.

There are several things to consider if you are looking for a new site.

- How close is it to town? You will need to balance the darkness of the sky to the length of the drive.
- How many people will the site accommodate? If it is just a site for you and a few friends, then a small clearing will do. If you are looking for a place for your astronomy club of 50 members, then you will need more room.
- Are there amenities nearby? A campground with a public toilet is great, but it may attract lots of other folks with torches ablaze. Americans – a torch is a flashlight; I just know you were picturing Frankenstein movies. I generally pick a spot that has some type of small town nearby so that I can purchase supplies during the day. Also, I have never needed it, but a nearby hospital could prove handy.
- How much of the sky can I see? There are many locations that look great on a map, but just aren't worth it once you arrive at the site. One thing that no map will show is the number and height of the trees at the spot. You just have to invest in the fuel and time to go see what a site looks like. When A.J. Crayon and I were looking for a site in the Central Mountains of Arizona it just took time. We left early in the day (that is 10 AM for me) and spent our time driving down roads and determining how much of the sky could be seen through the trees. A tip about doing that – take good notes. Our notes proved to be invaluable as a memory aid; they told us which turnoffs were good and which ones were poor.

The Saguaro Astronomy Club set up near the red rocks of Sedona, Arizona.

Jim Barclay at Maidenwell Observatory in Queensland, Australia. He has three 14 inch SCTs ready to show off those terrific southern skies.

My most expensive site is half a world away. I have been corresponding with an Australian named Jim Barclay for over 20 years. Jim and his wife Lynne have been nice enough to allow me to stay with them for a few weeks and observe the part of the sky that I cannot see from Arizona. Jim has also started a public viewing observatory near the little town of Maidenwell, Queensland. For a modest fee, you can have a guide to the southern skies. The Internet address is www.sbstars.com. Tell Jim I sent you.

Staying Warm

I can absolutely guarantee that if you are not warm enough while observing, your session will be cut short. Because you are standing still at the eyepiece, you are generating very little body heat and it is easy to get chilled. Being uncomfortable with the temperature will very quickly have you thinking about your body and not the celestial scene in the eyepiece.

Obviously, a heavy coat is a great place to start. Adding coveralls or a ski bib overall will keep your legs warm. Boots and two pair of socks are great for warm feet, one of the first places to get cold. Thin gloves or "rock climbers' gloves" with the fingertips exposed will allow you to write some notes on a chilly evening. A wool cap or a parka is also a big help; lots of heat can escape from your head. I use a muffler to keep warm air from leaking out from the collar of my coat.

There are two modern aids to keeping warm that I have found very handy. The editors of *Amateur Astronomy* magazine, Tom and Jeannie Clark, told me of a modern fiber used in socks; one brand name is "Therlo." The socks themselves are thin and made to be used with a thick woolen sock over them. Once I have the two pairs of socks on my feet and a good thick pair of boots, my feet are never cold and have not been cold since I started using them. Highly recommended.

The other recent development consists of heat packs. These small packets of activated charcoal will warm up when they are exposed to the air. They come in a variety of sizes, including some that are meant to be put into a boot. You just open the plastic wrap that covers the heat pack so that it is open to oxygen from the air and it starts to warm up. I keep them in my pockets and toss them into the sleeping bag when ready for bed – toasty. Rich Walker told me that he buys one that is meant as a back warmer and it keeps him warm through the night. My beautiful wife, Linda, is a nurse and she says that having a warmer over your kidneys will warm up all the blood in circulation through your body.

Car Camping

If you are lucky enough to live in a place far from the city lights, I envy you. However, most of us will have to camp out to enjoy a night under dark skies. That means we will have to think about car camping, as compared to hiking.

Obviously, the first thing to consider is the vehicle itself. I have owned two trucks with a bed long enough for me to sleep there when the telescope was set up. I lit-

Here is Ken Reeves using a set of ramps to get the mirror box of his 20 inch out of the back of his truck. Astronomers are notoriously innovative.

I just had to prove to you that it snows in Arizona. This is my Subaru Outback Wagon at my sister-in-law's cabin near the town of Heber. I was there writing this book, obviously not observing.

erally measured the length of the bed of my second truck while it was still on the car lot, just to make certain that the tube assembly of the 13 inch scope would fit in the back. I observed many years with my 13 inch Newtonian in a long tube; it also had a large Bigfoot mount, manufactured by Pierre Schwaar. I spent many weekends out with this setup and it allowed me to enjoy astronomy as a tourist for many years.

But things change. I was ready to move on to another telescope and vehicle. I purchased the Celestron Nexstar 11 about the same time as I bought a Subaru Outback Wagon. I find the combination excellent. The car is much easier to drive than my old trucks and the NS 11 provides excellent views of a wide variety of astronomical objects. My point is: consider both your telescope and your transportation when buying either. I understand that many of you have a family sedan of some type and that will not change soon. Do the best you can and realize that with a "low rider" sedan you will be limited to the type of roads you can travel.

Having a trustworthy vehicle is a must. Keep your vehicle in top condition, because being stranded far from civilization will make for a bad day. Keep a good spare tire handy; bring a blanket and some water if the worst happens.

Lots of Stuff to Take Out (Remembering it All)

I have been trying to get more organized, honest. One approach is to create boxes of various sizes that will carry all the parts needed to put together a telescope once at the site. A trip to your local hardware store will provide you with lots of choices for plastic boxes and crates. Right now I have created three boxes – one for the Nexstar 11, one for the 100 mm refractor and one for "every time out." So each scope has its own box and then the things that go every time are in the third crate. It seems to work so far.

You will definitely need a tool kit for all the types of parts you use: conventional wrenches, Allen wrenches, metric or British. I am in the horrific spot of having a combination of both sizes. The Celestron and most of its parts are British and the RFT (Rich Field Telescope) refractor and its fittings are metric. What a world, what a world!

There are two methods to having all the tools you will need:

1 Carry all the possible tools you need that fit any telescope you have ever owned and some that only fit your friends' scopes.
2 Determine which fasteners on your scope or accessories you would try and manipulate out in the field and only carry those tools.

Yes, I plan on having my tool box like that one of these days. But, right now, I am in the middle of writing this book. So I carry out a big tool box and have lots of tools and other parts stuffed into it. Having a big tool box does make you popular when something breaks on a friend's telescope.

If you showed up at my garage right now you would see a list on the wall. It reminds me of all the stuff that I need to take out observing. Please make one for yourself and check it each time before you leave. Arriving at your observing site and realizing that you have forgotten something vital to the operation of the telescope, or its operator, will make for a very bad night.

I have been with people who have forgotten eyepieces, warm clothes, telescope parts, food, water, counterweights and star charts. Some of those things you can work around and some you cannot. If you are observing alone, then you are really going to have problems. When I have something special, such as an observing list or ephemeris for a comet, then I will make certain that I don't forget it by putting it into the car on the night before I leave.

There was a time when I was a Boy Scout and their motto "Be Prepared" is still worthwhile.

Chapter 3

Using a Computer

There are many things that a good computer can do for you in the area of observing the deep sky. I am not going to spend time going over types of computers and their components. Any well-made modern computer will do. A faster processor, more memory and a larger hard drive are nice, but not essential to a computer that will be useful for astronomy.

Step into the time machine for 1985. I think I took this photo for insurance purposes; it is a shot of my "286" computer. It ran at the amazing speed of 10 MHz, had 128 kilobytes of memory and a 20 megabyte hard drive. If I remember correctly, it cost about 3000 US dollars. Running DOS 3.3, I was on top of the world.

Planetarium Programs

There are lots of planetarium programs out there. Basically, they all show the sky as it will look with a variety of instruments from the naked eye to a quite large telescope. My personal favorite is Sky Map Pro by Chris Marriott. As many of these programs do, it contains a vast wealth of information about stars, galaxies and nebulae. The time of day or night, the day of the year and the size of your telescope can be selected to give a display unique to your electronic observations. Then you can make a printout to provide your own personalized star charts. Also, there is a Yahoo group that chats about Sky Map Pro. If you join the group you can have your questions answered by other members, or often by the author himself.

If you have never used one of these programs then you can get your feet wet with no cost to you by using freeware. In my opinion, the best of these free planetaria is HNSky. It is very powerful and still costs nothing; it is just a programming exercise for its author, Han Klein.

If you are just generally trying to find your way around and don't wish to be distracted by detail then try StarCalc. It is simple to use and shows enough of the sky to get started. Search the Internet with any of the search engines and they will steer you to the web sites associated with these programs.

Specialty Programs

There are several types of other computer programs that are helpful to astronomers. I use a program that shows moonrise and moonset for a selected evening. Many of the planetarium programs contain this information, but I find it helpful to have a quickly available small program to tell me the circumstances of a particular night.

Note Keeping

This is the other task that I do so often with the computer. In the field, I write down my notes and maybe do a drawing the old-fashioned way – on paper. Once I return home I type in the notes and file them into the correct constellation file. It is the way I have been organizing my observing sessions for decades. If another method works for you, that is great. Several of the planetaria programs have the ability to do note taking built into them. You could call up the map that includes an object and see your previous notes displayed.

Once I have electronically entered my observations, then I can do a wide variety of things with them. I have been writing articles for both *Amateur Astronomy* magazine and the web site cloudynights.com. Having my notes easily at hand and ready to use in those articles makes them much easier to compose. Also, if an email group that I participate in is discussing a particular object, I can provide my observations and other notes by just using the standard cut-and-paste commands.

Laptop vs. Desktop

This is an interesting debate and I have yet to come to any conclusion about which is better. Right now, I do not take my laptop into the field. There are two reasons. First, I wonder about trusting the technology when the computer will be outside on a chilly night. If the battery dies and I have no backup, then I am back to entering notes and drawings on paper. Second, I have yet to see a display screen on a laptop that does not ruin my night vision by being too bright. If there is enough red plastic over the screen to darken it significantly, then the screen is too dim for my old eyes to read. I understand that having a laptop in the field gives you lots of flexibility and having all my notes and charts at hand sounds great. I just believe that right now I don't trust the system enough to rely on it.

There is something else that is less tangible and more spiritual. I have a file cabinet that contains my observations in manila folders. I can go to that folder and pull out the actual piece of paper that I used that night. I can view the drawings and notes in their raw form. That provides me a real link to past observing sessions that an electronic file cannot.

Web Sites

It is certainly easy to get lost on the Internet. There is much to see in the electronic gabfest. All I can do is provide some of my favorites and hope that they prove useful to you also. This is not in any order.

www.saguaroastro.org

This is the web site of the Saguaro Astronomy Club. It contains a free download of the SAC database, lots of information on the brightest 10,000 deep sky objects around the sky. My notes are also available.

www.cloudynights.com

Astronomics is nice enough to sponsor this web site. Obviously they wish to sell you something, but much of the information is free. I have been writing deep sky articles here for some time and the most recent and past articles are available.

http://www.relex.ru/~zalex/main.htm

This is the site for Starcalc, a simple planetarium program.

http://www.hnsky.org/software.htm

This is the site for HNSky, an excellent freeware planetarium program.

http://www.blackskies.org/links.html

This is Doug Snyder's nebula web site; it is a direct path to the "links" page. It contains lots of great URL links to astronomy information of all types.

http://www.ngcic.org

This is Bob Erdmann's NGC Project web site that contains information on the best positions and details on the NGC catalog. There is also a great "links" area here.

http://www.messier45.com

This is the Deep Sky Browser, a very useful resource that will allow you to create many types of observing lists for all types of deep sky objects. You can keep busy here for many lifetimes.

Chapter 4

A Fun Night Under the Stars

In this chapter I will explore what it takes to create a fun night out observing for me. I hope that you will see some patterns that you can add to your observing sessions.

Observing List

I always make an observing list of some type. It seems to me that time under dark skies is precious and I need to be careful about wasting it. Try and add a few new objects, ones you have never seen before, to your list. If you are going to observe an old favorite then give yourself a note about some detail within that object you have never seen before. Maybe look up a type of object you have never seen: a supernova remnant, for instance.

Take on a big list that will provide you with objects for years to come. The Messier list is the most famous, but the Herschel 400 has some bright and famous members. If you are looking for a big challenge then *Burnham's Celestial Handbook* contains a long list of objects that will keep you busy for a decade or two. I completed observing Burnham's entire list, at least the portions that I could reach from Arizona, about four years ago. It was a great feeling to know that I had taken on a big challenge and completed it. The Saguaro Astronomy Club database lists the brightest 10,000 objects around the entire sky. If you finish that I will be interested in the story of how you did them all.

If you don't wish to make up your own list, then every major astronomy magazine has a deep sky article in it. Just pick up the latest issue and observe along with the author for that month's listing. Obviously, I planned for you to observe the objects in this book, so there is information about each of the objects in the four chapters on observing by the seasons. The appendix in the back is a large sheet of information on many different nebulae around the sky.

Getting Ready to Go

I know that it might seem trivial, but a nap is a great observing technique. If you can get a little rest before going out observing, then do so. If it is difficult to get an afternoon nap at your house, then leave early, set up the scope, and lie down in the back of the car. You will really be ready by the time it is dark.

After loading up the car and checking the list to make certain that I have everything, I am ready to go. It really does pass the time faster to be going out to the site with a friend. If it is both of you in one vehicle or forming a caravan of vehicles, there is safety in numbers. We often chat on CB radios on the way out and back. It makes the time go fast and does provide a margin of safety as we can alert other people to road hazards. In the US the truckers are usually on Channel 19, and Channel 9 is for real emergencies, so choose a channel far from those.

We generally get on the road early enough in the afternoon so that we are at the site before sunset. This allows for an hour of setup time during twilight.

At the Site

We generally park the vehicles in a side-by-side configuration; this seems to give the most room for everyone to work from the end of their vehicle. I have a folding table that has proven very handy for setting out all the accessories one needs to use a telescope. If there is the possibility of dew getting everything wet, I will bring some towels to cover it all. There are coverings made specifically for many different types of telescopes, but I have not purchased one yet.

This setup allows me to spend the least amount of effort moving from the scope to the notes and star charts in the back of the car. Usually, the thing that wears out first is your feet. So, again, a little rest is a great observing technique. One of the folks in the group will call out "Break in 10 minutes" and we will get out the folding chairs and gather round the table to eat, rest, chat about what we are observing, and generally get off our collective feet.

Get into your cool-weather gear early. I have found that if I allow myself to get cold, it is difficult to warm back up and get comfortable. I put on a layer at the earliest sign that it is getting chilly: that is, usually in twilight.

A snack and some water is also a good observing technique. If you are going to be moving about for hours you can get cold and tired just because you don't have enough fuel on board. So, during one of those breaks, eat a little something and drink some water also.

If you are driving back into town that evening, it is imperative that you leave early enough to make the journey home a safe one. Falling asleep at the wheel is a

The Saguaro Astronomy Club, ready to observe. We line up the cars so that the most people and telescopes can get set up. Also, you can use the tailgate of a truck as a work space.

sad way to damage both you and your telescope. I know that it is difficult to dis-assemble the scope when the stars are bright overhead, but it can prevent a tragedy. Again, talking on the CB radio can keep you awake. You do have to hold up your end of the conversation.

Back Home

Now, as I unpack the telescope and other equipment, I set aside anything that needs to be cleaned or repaired. It is aggravating to go observing next time only to find that you have not fixed the accessory that was broken last time.

The next task is entering my notes into the computer. I try and do that within a day or so, that way I can remember what I saw and provide good electronic notes about what it looked like in the eyepiece. Then I file the paper notes into my filing system.

The Trip to "Five Mile Meadow"

In 2004 the Saguaro Astronomy Club had one of its best outings at a spot we call Five Mile Meadow, because it is a meadow and it is 5 miles off the paved road. This site is at high altitude – 6800 feet or about 2100 meters. We had previously scouted the site to make certain that a family sedan could make the trip without damage to the undercarriage. I packed up the Subaru Outback Wagon and checked my list to make certain that I had everything I needed for a fun weekend at a remote site. A.J., David, and I drove up on Friday afternoon without incident, gabbing on the CB radio about everything from the weather for the weekend to what new accessories we planned to use tonight. We stopped at the ranger station to let them know where we would be, always a good idea.

This is Five Mile Meadow, about 30 miles from Flagstaff. The trees get you some shade during the afternoon and the skies are excellent at night.

As we got to the site we realized this was going to be a great night. There were no clouds in the sky, and as we set up the scopes lots of other observers arrived. By twilight there were 22 vehicles and plenty of scopes and observers to soak up the starlight. We had a great weekend and we rated the transparency of the first night as 10 out of 10: as good as it gets. The Saturday night was 8 out of 10, still excellent. The Milky Way was wide and bright. I used my 100 mm (4 inch) RFT refractor to go after dark nebulae in Scorpius and Ophiuchus.

We did several things that made a big difference: we dressed warm. It was 100 degrees F (41 degrees C) when we left Phoenix on the desert floor; by 2 AM it was 28 deg F (−4 degrees C). I promise that is cold to a bunch of desert dwellers. So having all the cold-weather gear was important. Also we had done our homework; we had found a good site and created observing lists so we were ready to go when the weather was good on a new moon weekend. Some of this is luck (the weather), some was preparation on our part.

We did leave out one thing – insects. There are little flesh-eating bugs in that meadow and I, along with others, wound up with little red bite marks on our ankles. So next year we will stock up on insect repellant before leaving. One more thing to add to the list. It never ends. Be Prepared.

Chapter 5

Improving your Skills

As this is called the "Advanced Observer" series of books, I believe that my readers are willing to spent some time and effort acquiring the skills needed to become an avid deep sky observer. So here are some things that you need to know to become a skilled observer.

Getting and Staying Dark-Adapted

The human eye is a marvelous detector of light. During the day, it sees amazing detail in objects far and near. Once night falls, the eye and brain work together to provide the opportunity for your eye to see quite a lot in surroundings that are approximately 10,000 times fainter than full daylight. That is assuming the Moon is below the horizon. During a day when you are planning to observe that evening, wear sunglasses when outside. It seems to me that the dazzle of bright daylight makes your eye take extra time to get fully dark-adapted.

You don't have to "do" anything to have this physical and chemical reaction start to happen. You do have to protect it once you are dark-adapted. A single burst of white light can mean that it will take 30 minutes or so for your eye to return to its previous level of adaptation. So don't turn on the lights unless it is an emergency and don't go out observing with people who don't have good discipline about white light. It is easy – use dim red torches, turn off the door-activated lights in your car and make certain that the other observers around you do the same. Again, I said a dim red flashlight; I put some masking tape over the front of my flashlight to make certain that it provides a faint, diffuse light. I recommend it.

A.J. and I have found that a dark cloth over your head while at the eyepiece allows you to see to a deeper limiting magnitude. I also see more detail within the object I am observing when I am using the "monk's hood." I have a black cloth, A.J. uses a dark towel, but either way it makes you concentrate on the field of view and also blocks off external light, even at a dark site.

Learning to Really Observe

You have spent all this time, money and energy to get out under dark skies and provide yourself a good telescope and a dark-adapted eye. So don't just glance at that presentation of the marvelous Universe that is in the field of view. Spend some time really observing that nebula, cluster or galaxy. Use high and low power to get the complete view of the object and the field around it. Try a variety of filters; we will get to those in detail in the next chapter.

Try averted vision. It turns out that the most sensitive part of your vision is not dead center; it is off to the side. In the Navy we learned to "look above the horizon" when looking for distant ships at sea. I recommend the same technique to you. Look just above the object in the field of view and you will see fainter objects and more detail in brighter objects.

Some years ago I realized that the notes I had for some of the brightest and most famous objects turned out to be nothing more than "Wow." Therefore I created for myself the "Bright Object Project." I started with the Messier list and added the best of the NGC and created an observing list. Then I made myself really spend some time with each of these goodies until I had a good observation that really defined what I could see. I used binoculars, my finder scope, a variety of filters and magnification. It left me realizing that there was more to see if I spent the time to really observe and not just glance into the eyepiece.

I can't say enough about taking notes. My notes are the trophy that I return home with after a night of observing. I often return to my notes and read what I have seen previously. Having my notes available in electronic text makes providing my observations easier when I am chatting on the Internet. Obviously, this also makes writing this book easier. It isn't easy – but years of entering my notes into the computer do make it easier. I keep my notes as constellation files and, as you will see, that is how I will present them in the observing portion of this book.

Get Comfortable

I truly believe that you see more when you are comfortable. I try to sit down when I observe and that is an advantage of the Nexstar 11 and an adjustable-height chair. It is a joy to sit down and not have to worry about falling down. If you are a big

Sitting at the eyepiece provides a comfortable view.

Newtonian owner then consider your ladder carefully. A ladder with a large area on each rung is much more comfortable than a standard ladder. You want more than just a narrow area that digs into your instep; a large step allows your entire foot to stand comfortably on the ladder. One of the reasons that I owned the 13 inch f/5.6 Newtonian is that it did not need a ladder for most observing positions. When I was looking near the zenith I needed a short step, so I carried a wide board that allowed my whole foot to stand on that 2X10 (5 cm by 25 cm) inch board and be comfortable. A.J. calls his a "human shim."

Knowing the Magnitude of an Object in your Telescope

The magnitude system is not linear. That means that the values given would not form a straight line if plotted on a graph. This all started because your eye is not a linear detector of light. If one star appears twice as bright as another it is giving off about 2.5 times as much light as the fainter star. Also, a low number in the magnitude system is a bright star and higher numbers are faint stars. I keep this straight by remembering that a first-magnitude star is a "first-class" star; so it obviously seems brighter than a second-magnitude (or second-class) star.

Let me give you an example using nebulae. The famous Ring Nebula (M 57) in Lyra has a magnitude of 9.0 and it is easy to spot in both the Nexstar 11 and the 4 inch refractor. NGC 2346 is a planetary nebula in Monoceros and it has a magnitude of 12.5. I cannot make it out in the 4 inch, but it is pretty bright in the NS 11. After several years of experience with the 11 inch I can say that objects fainter than 14th magnitude are quite difficult with that telescope.

You need to determine the magnitude limit of your telescope. Most electronic planetarium programs will give you the magnitudes of the field stars in a particular area. Find a location that has a variety of magnitudes on the star chart and use it to determine the limiting magnitude. The best way is to go draw that field carefully and then compare it to the chart. There are also charts from variable star observers that are available on the Internet. You can use those fields as well.

There is one more complication here. It is called surface brightness. When we are discussing the brightness of an extended object and not a star, the nebula is spreading its light over a sizeable area and the star is a point. So the Ring Nebula may have the brightness of a ninth magnitude star, but that comes from an object that is much larger that the disk a star presents to the observer's eye. Because the star is more concentrated it is going to be easier to see than a nebula of the same magnitude. Another excellent example is the NGC 7293, the Helix Nebula in Aquarius. This object is given a total magnitude of 6.0, but it is spread over a large part of the sky. So, even though a person with average eyesight can see a star of 6th magnitude, there is no chance of seeing the Helix Nebula without at least a pair of binoculars.

Knowing the Size of an Object in your Telescope

Now that we have some idea of the brightness an object will present at the eyepiece, let's move on to size. There are three values associated with the size of an astronomical object. The largest is degrees: one degree is one 360th of a circle. For instance, at their largest dimension, the Lagoon Nebula and the Eta Carina Nebula are about 1 degree in size. Moving to small angular measurements we get to an arc minute, that is one 60th (1/60) of 1 degree. The Ring Nebula is 1 arc minute in size. Lower still is the arc second, that is one 60th of an arc minute. On a night of good seeing, the tiny disk displayed by a star at high magnification is about 1 arc second in size. Many smaller planetary nebulae are only 2 to 5 arc seconds in size.

An excellent object to observe differences in size is the star cluster M 46 in Puppis. The star cluster has a size of 27 arc minutes. I realize that sometimes it is difficult to tell when the Milky Way blends into a star cluster, but do your best. Then notice that there is a small nebula at the edge of the cluster. That is NGC 2438 and its size is given at 65 arc seconds. Often the given size of a nebula is larger than it will appear in your telescope, because it was measured from a long-exposure photograph or CCD image. Give yourself a moment to compare some given sizes from an astronomical reference versus what you see in the eyepiece. That way you can get a feel for the units associated with angular size and how they appear with a given magnification in your telescope.

Here is NGC 2438 and M 46 from Chris Schur with a 12.5 inch f/5 Newtonian.

Knowing Orientation or Position Angle in your Telescope

The next thing that will be helpful to you is acquiring some knowledge about the cardinal directions in the eyepiece of your scope. This way, as you are observing and trying to use a star chart, you will know which directions are which.

Just push the scope toward the north (at Polaris) and the side of the field of view where stars are entering the field is north. Those of you who live south of the equator can push the end of the tube toward Octans. OK, underneath the Southern Cross will be close enough. Then push the end of the scope toward the eastern horizon and the stars in the view will enter on the eastern side. That one works regardless of where you are on the Earth. If you have an equatorial mount, this is easy; the mount is already set up to move in the cardinal directions. With an alt-az mount you will need to just guess the best you can. As the alt-az system moves around the sky, the compass directions in the field of view change with the angle above the horizon.

Once you know the orientation of the field, you can do two things with that information. As you read through my notes you will see that I often give a position angle (PA) of elongation for some of these objects. You can make your own estimate and see if it agrees with me or other observers. The position angle values start at north with 0 degrees, then clockwise to east at 90 degrees, on to south at 180 degrees and west at 270 degrees.

The second thing you can do with this information is compare it to a star chart. If you are pointing the scope at a known star and wish to move to the southeast from there, you will need to know the cardinal directions in the eyepiece so you can move in that direction.

I promise there is some fun to be had here. It seems like a big task if you have not been doing any of this in the past. However, I guarantee that you will start to enjoy your observing more if you learn to use your telescope and take good notes. One of the real joys of this hobby for me is the opportunity to read through my notes and re-live observing sessions from years ago. Remembering times past can trigger a reason to observe something again or push you to try some objects that you missed the last time out. It is supposed to be fun, honest.

Chapter 6

Nebula Knowledge

Now that you have a pretty good idea about what to do to become an expert deep sky observer, the rest of this book will be specifically about nebulae. The word nebula is Latin for "cloud." In the early history of astronomy, any fuzzy object in the telescope was called a "nebula." As time went on and we knew more about the Universe it was realized that most of these objects called nebulae are galaxies: far distant star cities that look fuzzy just because they are so very far away. With that knowledge, astronomers only continued to use the word nebula to designate gas and dust clouds within the interstellar medium.

So let's start with a discussion of the five types of nebulae and lead that into a discussion of wavelengths of light, spectra and using filters to see the most in these faint glowing clouds among the stars.

Emission Nebulae

An emission nebula is a cloud of gas that emits photons of light. Many nebulae have stars embedded in them and because stars are born within nebulae, those brand-new stars are often very hot. Our Sun is about 6000 degrees C and therefore it is yellow. These hot O and B type stars are 9000 to 12,000 degrees C and therefore give off lots of ultraviolet radiation. This is the same type of radiation that is energetic enough to sunburn your skin.

This ultraviolet radiation in the nebula also delivers energy to the atoms within the cloud. It ionizes the atom, which means it delivers enough energy to knock electrons loose from the outer shells of that atom. When those electrons eventually recombine with the ion, they give off a photon of light at the moment of recombination. But not just any photon. The energy of the light emitted is at a very specific wavelength. That wavelength is determined by the type of atom involved. In other words, you can tell the chemistry of the cloud by the precise color of the light emitted by the atoms contained within the emission nebula. Not only is there a star within the nebula giving off light, the gas of the nebula itself is glowing as well.

Reflection Nebulae

Reflection nebulae are dusty and do not glow on their own. Nearby stars are giving off light and that light reflects off the dust and gas within the nebulae. We are just at the correct angle to see the reflection. Much as a cloud above the setting sun will reflect sunlight and glow brightly for half an hour or so after sunset, these clouds glow because we are at the correct angle to view them glowing. Because of the size

This photo of the Lagoon and Trifid nebula in Sagittarius is from Jon Christensen with a Takahashi refractor. The Lagoon is M 8 and it is the lower nebula; the Trifid is M 20 and it is the nebula nearest the top of the photo. The upper portion of the Trifid is a reflection nebula; the lower half of the Trifid and the entire Lagoon are emission nebulae.

and composition of the dust within these clouds, the light reflected from them is blue. This is the rarest type of nebula.

Planetary Nebulae

Planetaries have nothing to do with the planets. The first ones were discovered in the same time frame as the planets Uranus and Neptune. These outer planets appear like a small blue-green disk in the telescope. Well, some planetary nebulae are also a small blue-green disk in the telescope. The similar appearance yielded the name that persists to this day. This is far and away the most common form of nebulae. If you don't believe me, start up your electronic planetarium program and aim it at the middle of the Milky Way in Aquila, Cygnus, or Sagittarius. Now make certain that the magnitude limit for deep sky objects is, let's say, 15. Look at all the little planetary nebulae that appear on the screen of your computer. Many are small and faint, but they are there.

Planetary nebulae result at the end of the lifetime of stars like the Sun. Smaller stars do not explode at the end of their life; they rapidly puff away the outer shells of gas and dust into space at speeds of approximately 15 km/sec. That is about 25,000 miles/hr! What is left behind is a nebula of gas and dust and a small, hot star in the middle that used to be the core of the star. That white dwarf star is shining at very high temperatures and will provide the ultraviolet radiation that ionizes the gas so that it will glow. The life of a planetary nebula is short – less than 50,000 years – and therefore there are only about 1500 planetary nebulae known to exist within the range of today's professional telescopes.

It turns out that only very special stars form planetary nebulae, called asymptotic giant branch stars. They go through a phase in which they puff off a significant amount of their mass by giving off a huge stellar wind. The stellar wind from a planetary nebula is spewing 1/100,000th of a solar mass per year. That is about 600 trillion tons per second! The solar wind from our star is only one hundred trillionth of the mass of the Sun per year.

However, these winds are spherical and symmetrical. Everyone who has spent any time looking at the amazing images from the Hubble Space Telescope knows that the shapes of planetary nebulae come in an amazing variety of forms. Even an evening spent looking at just planetaries in a modest telescope will show you a ring, a dumbbell, a cat's eye and a fake Saturn. All these shapes can't come from a simple ball of dust and gas expanding out into the interstellar medium.

Towards the end of the 1970s three astronomers, Sun Kwok, Chris Purton and Pim FitzGerald, came up with an idea about how this myriad of shapes could form. It is called the "interacting winds theory." Their theory says that at the beginning of a planetary the asymptotic giant branch star is giving off the "slow" stellar wind at about 15 km/sec and it does so for hundreds of years. Then, a "fast" stellar wind begins to blow and overtake the particles in the slower wind. When they say fast, they mean 3000 km/sec, making these particles the fastest moving physical phenomena in the Universe: about one hundredth of light speed. When the fast wind catches the slow wind, the dust and gas interact and create the amazing diversity of shapes we see in planetary nebulae. The best proof is seen in the images of many planetaries: there is a bright inner section with a faint outer shell.

There is one more level of complexity when exploring planetary nebulae. Evidence is increasingly showing that the center of many planetary nebulae is populated by a double star. This fact is given credence by the amazing differences in form and structure seen in planetaries. As the two stars revolve about their center of mass, one is spewing off gas and dust like a garden sprinkler. Mathematical models with supercomputers are showing that the forms seen in these nebulae can be re-created by computer programs. Many of the most fascinating shapes involve two stars dancing about each other and seeding the Universe with these complex nebulae.

The intricate outer nebulosity is easy to see in this image of the Dumbbell Nebula (M 27) by Chris Schur with a 12.5 inch f/5 Newtonian.

Dark Nebulae

Dark nebulae are clouds of dust and some gas that are not at the correct angle to become a reflection nebula. They block off the light from more distant stars and therefore are seen as a "star void." Some are so good at blocking off starlight that they completely hide all the stars on the other side. Generally, the light absorption happens because the dark cloud contains carbon, just like the graphite in pencil lead. Also, these clouds are cold, because if they were hot they would glow and be an emission nebula. The temperature of these dark carbon clouds is on the order of −200 degrees C. That is cold! Obviously, these black clouds are most easy to see along the Milky Way, where there are plenty of shining stars to provide contrast with the dark cloud.

The next time you are lucky enough to be under truly dark skies I hope that the Milky Way is above the horizon. If that happens to you next time you are out observing, notice that the Milky Way is not a simple, smooth shape. There are bright and dark places all along its path through the sky. Most of the bright spots are star clusters and the dark areas are these carbon clouds called dark nebulae. One of the largest and most prominent of these areas starts in Cygnus then spreads through Aquila and into Scutum. It is called the Great Rift, with good reason.

Easily the most recognizable nickname for a dark nebula is the Horsehead (Barnard 33) in Orion. This is by Chris Schur with a 12.5 inch f/5 Newtonian.

Supernova Remnants

One of the most spectacular and violent events since the Big Bang is a supernova explosion. Stars that are much more massive that our Sun do not reach the end of their life quietly. As they go through their life, the huge stars build up layers of various elements within them, like the layers of an onion. Hydrogen is fused into helium, helium into lithium, then oxygen, nitrogen and on up the periodic table to heavier elements.

Once the element iron is reached, there is a major change in the fusion process. The balance between the outward energy from the fusion process and the inward pull from the gravity of this massive star is disturbed. To fuse elements heavier than iron demands a gigantic input of energy at the core. The only way to have that happen is for the star to collapse. This happens in about one second of time! That is an amazing fact if you really understand how large a 10 solar mass star must be. In that one second the outer layers fall in on the core and it compresses under the massive implosion. The core can only compress so much and eventually the star material rebounds and blows itself to pieces.

The core becomes a neutron star, created by the gigantic pressure at the center of the explosion. The pressure is enough to compress the electrons and protons that used to make up the stellar material. Now the negative electrons and positive protons are compressed, they join together and neutralize their charges to form billions upon billions of neutrons. The neutron star has a mass close to our Sun and yet it is about the size of a modern city! If a modern suburban house were compressed into neutron star material, it would be so small that it would take a powerful microscope to see it. The most famous neutron star is in the center of the Crab Nebula (M 1) in Taurus.

This type of supernova (a single massive star exploding) is given the designation Type II supernova. A Type I supernova is an explosion that is the result of a close orbiting binary star. One star is rather normal, except that material from that star is being pulled off its outer layers by the companion. The companion is a very compact white dwarf that is accreting (gravitationally attracting) the in-falling matter from the larger star in this binary pair. This cannot go on forever and eventually the dwarf star explodes from the weight of the in-falling material and blows both stars to shreds.

The material that used to be the outer layers of these stars is blown into space. An estimate of the amount of material is that most supernovae remnants contain an amount of dust and gas equal to the mass of our Sun. Very hot stars near the explosion have plenty of energy to ionize those outer layers as they speed away from the explosion. And so the supernova remnant is born. Many of them look exactly like what they are: the ejected material from an exploding star.

This is how the heavy elements that used to be part of the layers within the star are recycled into the interstellar medium, eventually to make our Earth and your body.

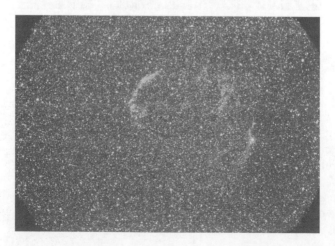

Chris Schur used his 8 inch Schmidt camera to photograph the Veil Nebula in Cygnus, the most obvious supernova remnant in the sky.

Wavelengths of Light

In what is one of the most bizarre revelations in modern physics, light can seem to be a particle or a wave. Depending on how you create the experiment, the radiation you are testing can exhibit either characteristic.

The beautiful colors in a rainbow provide a spectacular way in which light demonstrates that it is a wave. The droplets of rain within the clouds refract and reflect the light inside the raindrop and from that process the rainbow is revealed. A crystal prism can create the same effect. If you are wearing a pair of blue jeans, the dye in that fiber is absorbing all the colors other than blue and reflecting the blue light that you see.

Light on Earth can also exhibit particle characteristics. Every light-activated electronic circuit works because the photons of light are knocking electrons out of orbit and creating current flow within the circuit. The reason the streetlights come on automatically in twilight is the photo detector circuit that controls the light. Also, every digital camera collects photons to create the image it stores into computer memory.

For this discussion, I will only talk about light as a waveform from here on. The wavelength of a light beam is the distance from one peak of the wave to the next peak. It is the most important characteristic of the light from the stars and nebulae we will observe. Blue light has a shorter wavelength than a red light. Using sound waves as an example, a bass note from a drum or an organ has very different characteristics than a treble note from a saxophone or a flute. Even with a blindfold in place, we can tell the difference in what instrument created that sound.

Today the wavelength of a light beam is measured in nanometers (nm); this is one billionth of a meter. The dark-adapted human eye can see light waves from 400 nm to 620 nm. The blue end of your vision is around 400 nm and at 620 nm you are just seeing into the beginning of the red part of the spectrum. During bright daylight the red end of your vision extends to about 750 nm.

Astronomers have gotten very good at determining a lot of information about the stars and nebulae from the light they receive from these amazingly distant objects. After all, it is the only way they will ever get any information from a nebula that is 1000 light years away. No other form of communication is available to us.

One piece of information they can determine is the temperature. Heating a piece of steel in a flame, first it glows red, then orange, yellow, and finally blue-white as it gets hotter. The stars provide the same information. Red and orange stars glow at low temperatures. Betelgeuse and Antares are cool compared to the Sun. Our yellow Sun falls in the middle of the temperature range. Vega and Deneb are hot, bluish stars. By measuring the exact color and matching that to a laboratory experiment on Earth, a scientist can tell the exact temperature of a star.

Spectra

The science of astronomy really started to get on firm experimental footing once it was realized that the spectrum of a star or nebula provided so much information about that object. This started when the art of making glass got good enough to provide pure crystal prisms. The experiments of Kirchoff and Bunsen showed that there are three types of spectra, depending on the circumstance of the chemicals being tested.

The first type is a continuous spectrum. This is the complete rainbow of all colors with no breaks between them. Many materials when heated to incandescence will emit a continuous spectrum. The peak of this emission will allow you to determine the temperature of the material.

The second type of spectrum is an emission spectrum. A low-pressure, but high-temperature gas will emit light as the electrons in orbit around the nucleus give off photons of energy. The most important finding was that the emitted light is given off in specific wavelengths. The color or wavelength of the emission will inform the observer of the chemical makeup of the gas. This type of spectrum is seen as narrow bright lines that shine in the specific color given by the chemistry of the emitting gas. Glowing nebulae are obviously an example here, and so are fluorescent lights. If you hold up a prism at the correct angle you can see the emission lines of neon as given off by the light tube.

The third type of spectrum is an absorption spectrum. Very quickly it was seen that the rainbow of colors in sunlight contains dark lines across the solar rainbow spectrum. Again, atomic structure solved the mystery. The electrons in orbit can also take on, or absorb, energy from their surroundings. However, this can only be done at specific wavelengths. As the electron moves to a higher orbit with more energy it can only do this in specific step sizes; no partial steps are allowed. The atom removes energy from the continuous spectrum in precise wavelengths and these show up as dark lines across the rainbow of colors.

The Doppler effect provides information about the movement of the star or nebula. The wavelength of the light changes as the object moves toward or away from the observer. Nebulae moving away from Earth are stretching out the wavelengths of light and shift the color toward the red. Objects moving toward the observer are blue-shifted.

Therefore, the brightness, position, thickness and even tilt of these lines within the spectra can provide a wealth of information about the chemical composition, density, rate of movement and spin rate of the nebula you are exploring. One of the main reasons that astronomers are pushing to build larger and larger telescopes is to be able to capture fainter objects and acquire their spectra. This is how we know so much about the Universe.

Filters – What They Do and How to Use Them (See color insert)

All filters allow through some light and block off other parts of the spectrum. That is the reason that the information given above was not just fun to know. There are several sources of light that an observer of nebulae would like to remove from the field of view. Natural airglow is a faint glow from the atmosphere of the Earth. Manmade city lights are the worst at lowering the contrast between the background sky and the nebula you are trying to observe. Zodiacal light and starlight in general don't help either. So the trick is to allow the light from the nebula to make it through the filter and remove the glow from other sources.

Enter the modern "nebular filter." This is a carefully made piece of glass with a coating that has very carefully controlled optical characteristics. The good news is

Most often these nebular filters are used by threading them into the bottom of an eyepiece and inserting the whole assembly into the focuser tube.

that with today's layering techniques a filter can be made that is very specific to the type of object you are trying to observe. There are three general types of these filters. There are a series of color charts in the center of this book that give the wavelengths of light that are allowed through the coating of each of the filters I will discuss. This area of the visible spectrum which is allowed through the filter is called the "passband." Different passbands make each filter more useful for differing nebulae.

The broadband filters allow through the most light, hence the name. They block off several wavelengths that are emitted by sodium vapor and mercury vapor streetlights. They are most useful when your observing site is light polluted. My Orion Skyglow filter does a good job from my backyard. It increases the amount of the Orion Nebula that I can see and enhances the Lagoon and Trifid nebulae quite a lot. The best examples are the Lumicon Deep Sky and Orion Skyglow filters. However, if I drive out to moderately dark skies, it is less useful. Sorry, but the best observations are always provided by the "gasoline filter." (That joke will be "petrol filter" to my Australian friends.) The further you get from the city, the better the view.

Narrowband filters are tuned to pass several bright emission areas and more aggressively block the emission from streetlights. The most famous of these is the Lumicon Ultra High Contrast (UHC) filter. Similar is the Meade Narrowband filter. There are three main emission wavelengths from nebulae that are passed by these filters: Hydrogen Beta (Hβ) at 486 nm, Oxygen III (OIII) at 496 and 501 nm and Hydrogen Alpha (Hα) at 656 nm. I find these filters the most useful on the largest number of nebulae. They darken the background sky very nicely and allow the light from the nebula to dominate the field of view in the telescope. As you read through the observations that make up the main body of this book, you will find many places where I got the best view of the object using the UHC filter.

Line filters are the third type of filter. They pass the narrowest portion of the light from a nebula, centered up on only one or two lines of emission. Easily the most famous and useful of these is the Oxygen III (pronounced oxygen three) filter. When an oxygen atom has all 16 electrons then it is Oxygen I. If a high-energy ultraviolet light strikes the oxygen it will ionize it by knocking two electrons away from the outer atomic shell. Now the oxygen ions are designated Oxygen III. When the two electrons return, the atom emits radiation in two narrow bands at 496 nm and 501 nm. This is in the wavelengths that your eye sees as green. The other line filter that is useful for some nebulae is the Hydrogen Beta filter. It passes an emission band at 486 nm: you would see that as blue-green or aqua. The temperature and density of the nebula must be at just the correct values to allow this band to be the majority of the output of the nebula. Those gas clouds are somewhat rare. But one very famous object is a dark nebula in front of a glow in this region. Because of this, these filters are often called "Horsehead filters." The famous outline of a horse's head in the constellation of Orion stands out nicely with this filter, if you have a big scope under dark sky to start with.

So, as the observer changes between the three types of filters, one of three choices is being made. Using a broadband filter will get rid of some of the light from bothersome streetlights and pass the most light onto the eye of the observer. It has the most effect under mild light pollution and it will work on galaxies and star clusters as well as nebulae. Inserting a narrowband filter will allow through several bands of light that are emitted by nebulae. In general, the Hydrogen Alpha, Hydrogen Beta and Oxygen III bands are passed through the filter. This means that a narrowband filter will provide increased contrast for many emission and planetary nebulae. A line filter will pass through only one type of radiation, generally one of the types mentioned above. The line filter is most severe, but can give very contrasty views of nebulae that emit in that wavelength.

The good news about these filters is that the background sky is black when they are in use. The bad news is that they only allow through a very small amount of light therefore the object must be pretty bright and the telescope must have a larger aperture to show any detail.

The color center section of this book contains some beautiful color images of nebulae, as well as color diagrams that provide information about the passband of the popular nebula filters. Notice that the passband of the DeepSky and UHC filters is wider than the OIII and H Beta filters. Also, they will show that amospheric natural airglow, Sodium (Na) and Mercury (Hg) street lights are outside the passband of all these filters, therefore this unwanted light is filtered out.

Chapter 7

Introduction to Observations

Up to here we have discussed observing nebulae in general. After this chapter the rest of the book will be about specific nebulae and what can be seen within them. I just want you to know that this is not a complete set of information on all the nebulae in the sky. The number of pages assigned to this task by my editor would not allow that. Besides, I would not want to try and write that book.

What I am going to try and do is provide you with specific examples of all the different types of nebulae discussed previously. I will let you know what I have seen of these objects and include some observations from myself and a few friends of mine.

My hope is that being able to read observations from astronomers who have been viewing the sky for many years will help you learn the word usage needed to report what you can see at the eyepiece. It is indeed like learning a new language. Day in and day out, we just don't speak of things being "irregularly round" or "somewhat brighter in the middle."

There are lots of observations of planetary nebulae, far more than any other type. Besides the fact that I enjoy observing them, planetaries are far and away the most prominent type of nebula. But please don't worry. I really like chasing bright and dark nebulae of all types and I have tried my best to provide a good cross-section of nebulae other than planetaries within my observations.

So my coaching is to make use of the information, tips and techniques in the first part of this book and go out and find some of the nebulae in the appendix. The appendix contains a listing of a wide variety of nebulae across the entire sky. If you decide to chase more of these fuzzy denizens of the deep sky, then the web sites will provide plenty of targets.

A recent trip to Australia allowed me to view some southern nebulae. The Appendix list is anything but complete, but it is a good start at what to view if you are a northerner making a trip to southern skies. It is worth the price of a plane ticket, I promise. At least my wife and the tax collector believe that it was worth it.

Chapter 8

Northern Autumn (Southern Spring) Nebulae

In Arizona this is a great time to be out observing. The heat of the summer has finally broken and the nights are comfortable and there is a lot to see. Right after sunset the summer goodies are available for viewing and if you stay up late you can get at Cassiopeia and Perseus.

Nebulae in Andromeda

NGC 7662 PLNNB AND 23 25.9 +42 32

E.E. Barnard made the discovery that the nucleus of this planetary nebula seems to vary considerably in brightness. Between 1897 and 1908 on nearly 80 dates his notes show a magnitude range from 12 to 16 for the central star. The reality of these changes has been questioned, however, by modern observers. C.R. O'Dell points out that the apparent brightness of a star surrounded by strong nebulosity is critically correlated by the seeing. "As the seeing varies, the ability to discern the star will change because of the superposition of the nebula, while nearby comparison stars will not be affected."

Using the 6 inch f/6 while checking a simple alt-az finding system using an HP 48 calculator program – it works! The seeing and transparency both rated the sky at 5 out of 10. Even with the low power of a 22 mm Panoptic eyepiece this object was recognized as a planetary nebula. Moving the magnification higher with the 8.8 mm shows it as bright, pretty large, and aqua in color. The central star is seen 10 percent of the time.

Using the 13 inch f/5.6 on a great night far from the city lights; rated at 8/10 for transparency and 7/10 for seeing. The nebula is a lovely blue-green disk at 100X. At 150X it is bright, pretty large and round. 330X provides a great view! Internal detail shows turbulent swirls, the central star comes and goes, showing itself as a stellar point about 30 percent of the time. Higher power shows some slight elongation and the NE side of the nebula is brighter. The blue color is not as obvious at high power.

This is an image of NGC 7662 by Chris Schur using a 12.5 inch f/5 Newtonian.

Aquarius Nebulae

NGC 7009 PLNNB AQR 21 04.2 –11 22

This planetary will show amazing detail within the disk on images from modern large telescopes, including the Hubble Space Telescope. Unfortunately, none of that is visible in even large amateur telescopes. But that does not mean that this fascinating object is not worth observing: quite the contrary. From a distance of 3000 light years there is much to see here.

The Saturn nebula has always fascinated me and my notes show that I have observed it often. Even with the 7 inch Maksutov in my backyard observatory it was easy to see. This is from north Phoenix, in the middle of a lot of light pollution. With a 10 mm eyepiece in the "Mak" this nebula shows a high surface brightness disk that is elongated 1.5X1. Averted vision makes it grow in size.

Observing far from the city lights really makes this bright planetary come to life. Using Pierre Schwaar's excellent 20 inch f/5 Newtonian also helps quite a bit. With a 12 mm eyepiece, the oval shape is obvious and there is a light glow that surrounds the nebula on all sides. This faint outer nebulosity is better seen with averted vision. The central star is easy with this aperture. The extensions, called ansae, are easy to see with direct vision and grow longer with averted vision. These exten-

Here is a shot of Pierre Schwaar's 20 inch f/5 Newtonian, mounted on his Giant Bigfoot mount. I did not get his face in the photo, so here is a shot of him drawing a sunspot using a projection eyepiece into his log book.

Here is a shot of Pierre Schwaar's 20 inch f/5 Newtonian, mounted on his Giant Bigfoot mount. I did not get his face in the photo, so here is a shot of him drawing a sunspot using a projection eyepiece into his log book.

This image of NGC 7009 was taken by Daniel Verschatse.

sions of the nebulosity are what give the Saturn nebula its name. What is most striking about this beautiful object is the fact that it is florescent aqua in color. The Saturn nebula really does glow in the dark.

NGC 7293 PLNNB AQR 22 29.6 –20 50

The Helix is the nearest planetary nebula to Earth, approximately 500 light years distant. Therefore the nebula is over one light year in size, making it larger than the outer reaches of the Solar System. Also, that distance makes the central star, at magnitude 13.5, about 1/12th the brightness of the Sun. Remember that the white dwarf stars within these planetaries are hot enough (75,000 degrees C and up!) to give off ultraviolet radiation that lights up the nebula, but they are the small cores of what used to be a rather normal star. The Helix is a large object but it is a low surface brightness nebula, so you need a large field of view to take it all in.

Using the best RFT I ever owned, a 6 inch f/6 Maksutov-Newtonian, the Helix Nebula is a great object. On a dark and clear night with a 14 mm eyepiece and a

This image of the Helix was taken by Jon Christensen with a Takahashi refractor.

UHC filter the central hole is easy. The overall view of this big planetary is bright, large, annular and round. Averted vision helps the view. Also, this object really responds to the UHC filter. At this power the nebula takes up about one third of the field of view. It is brighter on the east side. The central section is darker than the edges, but the middle is not completely free of nebulosity.

Viewing the Helix with the new Celestron Nexstar 11 and a 35 mm Panoptic is a real joy. The nebula is bright, very large, round and annular. The UHC filter really raises the contrast of this object and makes its famous "donut" shape unmistakable. It takes the low power and high contrast of the 35 mm Panoptic to appreciate this object.

With the 13 inch at 60X on an excellent night, this nebula is bright, large and round. There are five stars involved and a nice double star at the southern edge. The central area is darker than the annulus of nebulosity. There is no "halo" around the bright Helix of nebulosity. Moving up to 100X there are seven stars involved, and adding the UHC filter makes it very contrasty with a much darker middle. The filter reduces the number of stars to five, but provides a very nice view. It appears that 150X is the best view; the nebula takes up about 60 percent of the field of view. There are 11 stars involved, several very faint. The north side of the nebulosity is brightest. Adding the UHC at this power shows a very dark middle, but not completely free of nebulosity.

Cassiopeia Nebulae

NGC 281 CL+NB CAS 00 52.8 +56 37

This emission nebula is about 7000 light years distant and the multiple star in the center is Burnham 1. This O6 star is very hot and it ionizes the gas so that it glows. Brian Skiff says that the grouping of stars in the middle is typical of star clusters within nebulae in that it is probably only a few million years old – relatively young on the scale of the Universe.

In 4 inches (100 mm) of aperture and a 14 mm eyepiece, this nebula is just seen as a faint glow without a filter; there are six stars involved. Adding the UHC makes a big difference in the contrast of this nebula. There are only three stars involved

This image of NGC 281 was shot by Chris Schur with a 12.5 inch f/5 Newtonian.

with the filter in place, but the nebula now takes on the familiar "Pac-man" shape. It is still a low surface brightness nebula with the small scope, but the filter makes all the difference.

Using the Nexstar 11 and a 27 mm Panoptic eyepiece easily shows this nebula with 11 stars involved, including a triple star near the middle. The Pac-man is pretty bright, pretty large, very irregular in figure and a little brighter in the middle. Adding the 2 inch UHC makes this nebula much more contrasty. The shape is easy to see and there is a dark bay that forms the mouth.

NGC 896 BRTNB CAS 02 25.5 +62 01

This object is a low surface brightness nebula at the NW edge of a huge nebula complex in Cassiopeia. It includes NGC 896, IC 1795, and IC 1805.

Using the 4 inch (100 mm) f/6 refractor on a night rated at 5/10 for seeing and 6/10 for transparency; this nebulosity is tough in a small aperture. With a 27 mm Panoptic I saw it as very faint, pretty small, a little elongated. With no filter it is just barely seen, averted vision only. Going to the 22 mm Panoptic with the UHC filter is the best view, but it is still a small and low surface brightness nebula with an 11th magnitude star involved. The 8.8 mm Ultra Wide Angle (UWA) eyepiece is too much magnification for this object and it all but disappears.

With the 6 inch Maksutov-Newtonian at the 22 mm eyepiece on a mediocre night (S = 5, T = 6), it is seen as faint, pretty small and shows an irregular figure. This nebula is not much – just a glow around a pretty bright star.

Moving up to the 13 inch f/5.6 at 100X with the UHC filter it shows a curious footprint-shaped nebula. I found it hard to determine where IC 1795 stops and NGC 896 starts. This is a rich field; it is just filled up with faint stars.

This image of NGC 896 was taken by Sean and Renee Stecker/Adam Block/NOAO/AURA/ NSF at Kitt Peak; they were using a 20 inch (Ritchey-Chrétien) telescope, so the nebula is larger than the field of view of the CCD camera.

NGC 7635 BRTNB CAS 23 20.7 +61 11

In the 4 inch and a 14 mm eyepiece, the "Bubble Nebula" is a very faint round nebulosity surrounding two stars. Averted vision shows it off better, but it is still difficult. With the UHC filter the nebula is much easier to see, and now both stars have a glow around them that is about 1 arc minute in size.

In the Nexstar 11 with a 27 mm Panoptic eyepiece, there are two fuzzy stars, and the nebula is pretty faint, pretty small and round. Adding the 2 inch UHC makes the nebula about three times larger and the fainter of the two stars involved has a round, nebula with the star at the edge. It appears like "the Bubble."

This image of NGC 7635 was shot by Chris Schur with a 12.5 inch f/5 Newtonian.

IC 1747 PLNNB CAS 01 57.6 +63 19

This little planetary nebula is included so that you can practice on small planetaries. If you have never chased a small planetary nebula, then IC 1747 is a good place to start. It is bright enough to be seen in modest apertures, but you will need a good star chart if you are star hopping. If you are using a GOTO system, have it point at some other well-known objects to make certain that the system is finding deep sky objects accurately. Then point the scope at the area of this nebula and

This image of IC 1747 was taken by Adam Block/ NOAO/AURA/NSF.

make certain that you are at the correct location, put in an eyepiece that will provide 100X to 150X and you will see that one of the "stars" in the field is larger than the others; that is the planetary.

Using A.J. Crayon's 8 inch scope at a good site, I rated the night 6/10 for both seeing and transparency. This little nebula was spotted at 115X and showed a little elongation at 200X.

Using the Celestron Nexstar 11 on that same 6/10 night, this planetary was seen as a small non-stellar disk in a pretty rich field of view at 125X. Moving up to 320X with an 8.8 mm UWA eyepiece made it pretty bright and pretty small. It showed itself as a little elongated (1.2X1) disk which was very little darker in the middle. The large scope displayed a very light grey-green color.

IC 1848 BRTNB CAS 02 51.4 +60 25

This is a giant-sized emission nebula that is near IC 1805. There are also several open clusters in this area, a busy part of the sky.

Using the 4 inch f/6 RFT refractor with a 35 mm Panoptic provides a 3 degree field of view. There is a wide and bright double star that marks the western edge of this nebula. It is seen as fuzzy immediately in the 35 mm Panoptic with no filter. Most of the nebulosity is faint and with pretty low surface brightness, but it is visible right away on a good night. The central 50 percent of the glow is somewhat mottled in a rich field of view. Going to a 22 mm Panoptic raises the magnification, but the nebulosity is still easily seen. The southern edge is somewhat brighter and the UHC enhances that view. There are two clusters of stars involved: NGC 1848 shows four stars at a bright spot in the nebula and Collinder 34 is a large cluster with eight stars resolved.

This image of IC 1848(right side) is from Chris Schur with a Schmidt camera.

Cetus Nebulae

NGC 246 PLNNB CET 00 47.1 –11 52

This object can show you what a low surface brightness deep sky object will look like. Many planetaries are bright little dots; this one is not. The light from NGC 246 is spread over a large area and therefore is difficult to see in either a small telescope or on a night with poor transparency. NGC 246 is about 1500 light years distant.

This is one of the objects on which I have had the time to do a rather extensive test of nebular filters. This is one of the reasons mentioned earlier for joining an astronomy club – you get the chance to look through other people's telescopes. Then you can ask what eyepiece and filter they might be using. It gives you an opportunity to see what can be seen without having to buy that accessory and hope for the best.

With the 6 inch f/6 Maksutov-Newtonian and the 22 mm Panoptic eyepiece, NGC 246 is just seen as a faint fuzzy spot. Raising the power with an 8.8 mm shows that there are two stars involved within the nebulosity. It is very faint, pretty large for a planetary and is brightest on the north side. This is on a great night that I rated S = 6, T = 8.

On a night that was too poorly attended to be a star party (just the Nexstar 11 and me), this is an obvious nebula with stars involved. At 125X it was pretty bright, pretty large but not brighter in the middle. There are four stars involved, three of 10th or 11th magnitude; the faintest of these stars is 13th magnitude. This interesting nebula is elongated 1.5X1 with direct vision, but it appears almost round with averted vision. It shows a pretty low surface brightness, especially for a planetary nebula. Threading in the Deep Sky filter darkens the background, but it does get rid of some nebulosity; also the faintest star within the nebula is more difficult. Installing the OIII filter really darkens the field and gets rid of most stars in the field of view. The nebula is smaller but much more contrasty than without the filter. Adding the UHC filter also really darkens the field, not nearly as much as the OIII,

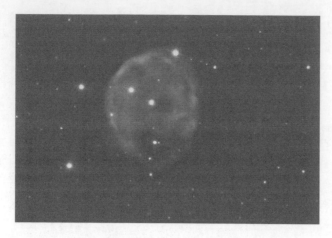

Image of NGC 246 by Jeff Cremer/Adam Block/ NOAO/AURA/NSF at Kitt Peak.

but quite a bit. I can only see the 13th-magnitude star about 10% of the time with the UHC. The nebula is much more apparent and shows some detail, the "right" side of the disk is much brighter and the middle is darker than without a filter. For me personally, I like the effect of the Deep Sky filter the best. Going to 200X also darkens the field and gets rid of some nebulosity. I can see the 13th-magnitude star now 100 percent of the time. The nebula is a donut, a bright rim with no middle; averted vision fills in the ring nicely with a nebulous glow. My observing buddy of many years, George deLange, said that this nebula "reminds me of a transparent balloon with a sprinkling of stars visible thru it."

Eridanus Nebulae

NGC 1535 PLNNB ERI 04 14.2 –12 44

Observing with a 6 inch f/6 Maksutov-Newtonian and a 22 mm Panoptic eyepiece, this planetary is pretty faint, small and shows a bright middle. Raising the power

This image of NGC 1535 was shot by Chris Schur with a 12.5 inch f/5 Newtonian.

with a 6.7 mm eyepiece shows it as pretty faint, pretty s\mall and having an approximately stellar nucleus. With averted vision it doubles in size. There is no color seen; it is a grey disk with a white stellar core.

Moving up to the Nexstar 11 with a 22 mm Panoptic, NGC 1535 is seen as bright, pretty large, round and has a central star that is quite easy to see, even at this low power. The nebula is a lovely light blue color that I find fascinating. Raising the power with the 14 mm shows the central star 100 percent of the time. There is some detail within the disk: a curved set of bright lines that create a "CBS eye" effect. The color is less prominent at higher power, but this is still an excellent planetary.

Pierre's big 20 inch at 280X provides an excellent view of inner detail; the CBS eye effect is easy. The central star is now very prominent. Overall, it shows a beautiful blue-green disk with several layers of brightness.

IC 2118 BRTNB ERI 05 04.5 –07 16

The Witch Head Nebula is a reflection nebula and a toughie! You will need a very dark night to be able to see this large and low surface brightness object. This dim streamer is a reflection nebula that is illuminated by Rigel. It is estimated to be 1000 light years distant. Once you have the telescope or binoculars pointed at the right position, try a break for a few minutes before straining your vision to see this nebula. I promise that a little time off your feet will help.

Using 10X50 binoculars on a night I rated 8/10, the Witchhead Nebula was extremely faint, very large and very elongated. On that same night I used the 6 inch f/6 telescope to view this difficult nebula. It was easier in the scope; it obviously provided more aperture. I installed several filters and the Deep Sky filter seemed to provide the best view. It darkened the field of view and still allowed enough of the light from the nebula to pass through the filter.

This is a photo of the Witchhead Nebula (IC 2118) from Chris Schur using an 8 inch Schmidt camera.

Grus Nebulae

IC 5148 PLNNB GRU 21 59.6 –39 23

There is a limited amount of time to view the constellation of Grus from Arizona. Grus is the Crane, the hoisting device, not the bird. It is well above the southern horizon for only a few months in autumn and early winter. I have not had the good fortune to travel to Australia during their spring and summer where the Crane lifts itself high overhead. Maybe I can use the royalty payments from this book to take my wife on a distant vacation.

The gas within this planetary nebula is moving faster than most, 53 kilometers per second. This expansion will eventually dissipate the gas and dust in the nebula out into interstellar space. The distance to this object is about 2900 light years.

In the 13 inch at 150X, this planetary is pretty faint, pretty large and little elongated (1.2X1). This nebula is annular – the dark hole can be seen with averted vision. Adding the UHC filter makes a big difference: the hole is much more obvious and it is a nice planetary. Get it while you can if you are observing from north of the equator.

This image of IC 5148 is by Jim Barclay with a 14 inch SCT.

Pegasus Nebulae

Jones 1 PLNNB PEG 23 35.9 +30 28

Jones 1 was found on a photographic patrol plate taken at Harvard University around 1940. Rebecca Jones discovered this object, which has received much attention recently from amateur observers. This planetary nebula is difficult, because of its low surface brightness, even on a good night. The overall magnitude from the SAC database is 12.7, but Dr. Harold Corwin's calculation shows that it has a surface brightness of 16.6 magnitudes per square arc minute. That is a faint object.

With a 10 inch f/5 RFT, Jones 1 was just barely seen. On a night I rated at 8/10 for transparency, I could just barely detect this object with a 14mm Ultra Wide Angle eyepiece and no filter. Adding the UHC filter made a big difference. The contrast of the nebula was increased very much and I could show this object to several other observers. This was before I owned an Oxygen III filter, and Ken Reeves was

This image of Jones 1 is by Adam Block/NOAO/AURA/NSF at Kitt Peak.

nice enough to loan me his. It also made this object easier and really blackened the background of the view.

The first time I observed this faint planetary was in the 13 inch on a good, but not great night. The evening was rated at 6/10. Using 100X, the only way to see this object was to install the UHC filter. It was very faint, very large and had an irregular shape. Jones 1 showed a "quarter moon" or "C" shape that got thicker with averted vision.

Perseus Nebulae

M 76 PLNNB PER 01 42.3 +51 35

This planetary nebula is the faintest Messier object in overall magnitude. However, it is quite small compared to a Messier galaxy, so it does have a high surface brightness and therefore is easy to pick out of the Perseus Milky Way. So a small nebula compresses its brightness into a smaller area and the surface brightness goes up. Distances to nebulae are difficult to derive, but 1700 light years is a good estimate for M 76.

Many years ago I realized that for many of the famous objects in the sky I only had a short set of notes that generally said something that added up to "Wow." So, as a long-term project, I took on making myself write down some comprehensive notes about what I was seeing while observing the bright and famous deep sky objects. Therefore, I will provide you with what I wrote about this object in the 13 inch on a good night. Even though I have observed this object in a variety of telescopes, I don't wish to bore you with too much detail. My point is that if you have a skinny notebook, then maybe you need to go back to some old favorites and really write about what can be seen. Most of the famous deep sky objects got that way because they show lots of detail on a sharp night.

At an observing spot about 60 miles from the lights of Phoenix, this little planetary can just be seen in the 11X80 finder; averted vision helps. In the 13 inch scope at 100X with the 22 mm Panoptic it is pretty bright, pretty large, rectangular in shape and elongated 2X1 in position angle of 30 degrees (north is 0 degrees, east is 90). It has a thin dark lane down the middle and there is a nice 9th-magnitude yellow star about 15 arc minutes east of the nebula. The nebula is a very light green color at low magnification and it is floating within a pretty rich field of view. Going to 220X shows why there are two NGC numbers associated with this object. The

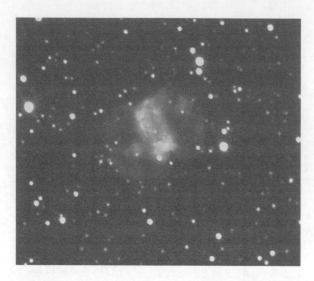

This is an image of M 76 from
Chris Schur with a 12.5 inch
f/5 Newtonian.

view is very different with direct and averted vision. Using direct vision shows two
distinct sections with a dark lane separating them. A faint extension of the nebula
to the south spoils a perfect rectangular shape. There are bright areas within the
nebulous glow to the NE and SW. These small bright spots are never stellar. The
"real" central star that illuminates the nebula is 15th magnitude and probably at
the limit of the 13 inch scope.

Using averted vision makes the Little Dumbbell almost twice as large by bring-
ing out a faint outer section, especially to the ENE. The "outer loops" are never held
with direct vision. The color of this nebula is now more grey than green. Using the
UHC and 330X brings out faint outer loops somewhat better than 220X. Going to
440X shows a faint star on the southern edge and a bright spot at the northern
edge that is stellar about 20 percent of the time. All of this is made more interesting to me when I remember the best theory about its shape is that we are seeing
a thick ring or torus that is edge-on to our view.

NGC 1491 BRTNB PER 04 03.3 +51 18

This nebula is included here to show you how the ability to observe the sky
increases with experience. When I first observed this nebula I called it "pretty faint"
and said "just barely noticed it." That observation was with my old 17.5 inch

This image of NGC 1491 is by
Chris Schur with a 12.5 inch f/5
Newtonian.

Five Mile Milky Way is by me from the Saguaro Astronomy Club dark site near the town of Happy Jack, Arizona. It is a 12 minute exposure with a 24 mm lens on Fuji 800 film.

M27-8H is an 8 hour (!) exposure of the Dumbbell Nebula in Vulpecula by Chris Schur with a 12 inch Newtonian.

Rho is an image with the 8 inch Schmidt camera of the region around Rho Ophiuchi. Antares and the globular cluster M 4 are at the bottom of the frame.

NGC 6357 is an 8 inch Schmidt photo of the area around this emission nebula near the "stinger" of Scorpius. NGC 6334 is the nebula at the bottom of the photo.

Bubble Neb is a photo from Richard Payne with his 9.25 inch Schmidt camera of the area around NGC 7635, the Bubble Nebula in Cassiopeia. M 52 is the prominent star cluster.

M8-M20 is from Jon Christensen and is what A.J. and I have called "Downtown Milky Way" for decades. It was taken with a Takahashi 210 Astrograph.

Keyhole is from Eddie Trimachi. An excellent closeup of the central area of the Eta Carinae Nebula, it is centered on the dark Keyhole region with Eta Carinae above it.

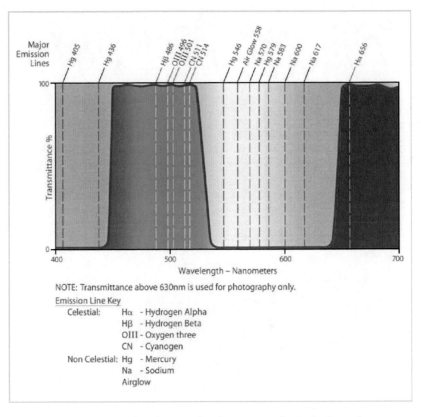

Lumicon Deep Sky Filter (Used with permission by Parks Optical)

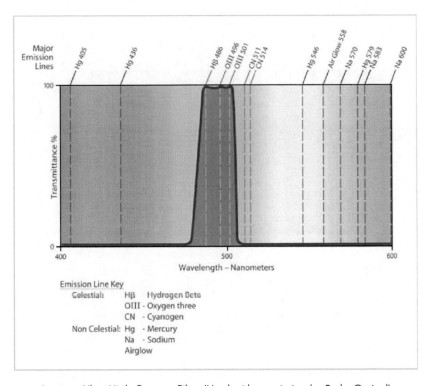

Lumicon Ultra High Contrast Filter (Used with permission by Parks Optical)

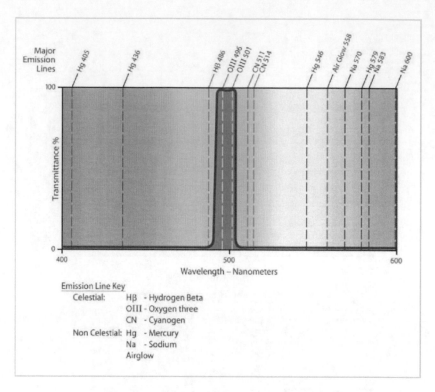

Lumicon Oxygen III Filter (Used with permission by Parks Optical)

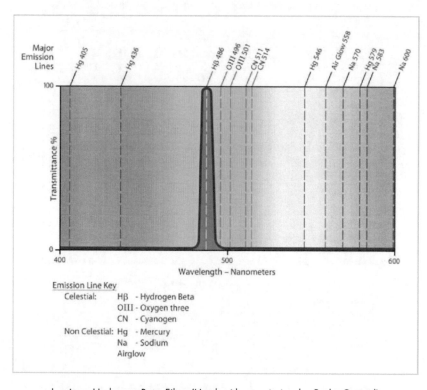

Lumicon Hydrogen-Beta Filter (Used with permission by Parks Optical)

(450 mm) Dobsonian when I had been observing for five years or so. Watch what happens 10 or 12 years down the road, when my observing skills were much better.

With the 13 inch at a mediocre site, at 100X this nebula is pretty bright, pretty small and a little brighter in the middle. There is a 10th-magnitude star on the eastern edge and four other 13th-magnitude stars involved. Averted vision doubles the size of the glow. Raising the power to 220X brings out two more stars for a total of seven. The UHC filter does not help the contrast of this rather low surface brightness nebula. It is triangular in shape both with and without the filter installed.

In an email message, Brian Skiff said that the brightest star involved is BD +50 886. This is a blue-white star when he determined its spectra. However, because it is shining through all that gas and dust in the nebula, it appears medium yellow.

NGC 1499 BRTNB PER 04 03.3 +36 25

The California Nebula is indeed shaped like that West Coast state. It seems to be illuminated by Xi Persei, a nearby bright star. My old 10X50 binoculars provided an excellent view on a dark night. The binoculars would show a faint, elongated glow that grew with averted vision.

The 4 inch (100 mm) RFT refractor was made for objects like the California Nebula. With a 27 mm Panoptic this famous nebula was pretty faint, very large, very elongated 3X1 and shows 11 stars involved within the glow. Adding the 2 inch UHC filter really enhances this huge nebula. The California shape is easy now. The nebulosity extends across the entire 2.5 degree field of view. Keeping Xi Persei out of the field really shows off the nebula at its best.

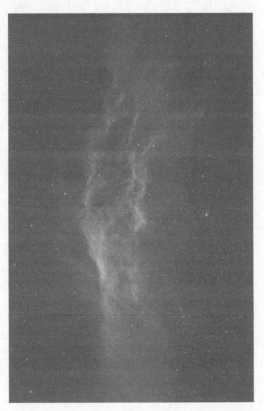

This image of NGC 1499 was taken by Richard Payne with a 300 mm lens.

NGC 1624 CL+NB PER 04 40.6 +50 28

This is one of the most distant nebulae that can be seen within our galaxy. The dust that makes up the Great Rift in the Milky Way blocks off much of our Galaxy when we are observing in the direction of Sagittarius, Scorpius, Crux, and Carina. However, when viewing toward Perseus we are looking away from the center of the Galaxy and can see further into space. So, NGC 1624 is about 20,000 light years distant from Earth, further than anything in the opposite direction.

With the 13 inch this nebula is pretty faint, pretty large and irregular in shape. The middle of this glow is brighter at 135X. There are seven stars involved in this nebula without the UHC filter. The filter helps the contrast somewhat.

This image of NGC 1624 is by Chris Schur with a 12.5 inch f/5 Newtonian.

IC 2003 PLNNB PER 03 56.4 +33 53

This is another very small planetary that will provide you an opportunity to see what most planetary nebulae are going to look like. They are generally small and just larger than the Airy disk of the stars in the field of view. This is also a great place to start learning the abbreviations that J.L.E. Dreyer adapted for use in the NGC and IC. Here is the description decoded: pB is pretty Bright, eS is extremely Small, lE ns is little Elongated north-south, *13 n 4″ is star of 13th magnitude that is north 4 arc seconds, and *12 sp 18″ is star of 12th magnitude that is south preceding (SW) 18 arc seconds. I know that getting started trying to figure out the code is not easy, but if you are observing NGC or IC objects then that information can be helpful.

I have only observed IC 2003 once and that was with my old 17.5 inch on a night I rated 7/10. I saw this nebula as pretty bright, pretty small, round and brighter in the middle. It was first noticed at 100X, but this little planetary stood out better at 220X. It is about five times the Airy disk, or stellar seeing disk, in size. There is a 13th-magnitude star to the SW about 20 arc seconds. IC 2003 is a nice lime green at all powers.

This is an image of IC 2003 by Chris Schur with a 12.5 inch f/5 Newtonian.

Chapter 9

Northern Winter (Southern Summer) Nebulae

I know that the winter sky is dominated by bright, easy to see, constellations. Please don't get so drawn to them that you don't spend some time observing in parts of the sky nearby Orion, Gemini, Canis Major, and Taurus. I promise that even though Monoceros, Puppis, and Pyxis are not as flashy as their companions, there is plenty to see in those parts of the sky. Also, during this part of the year our friends in the southern hemisphere are viewing the Magellanic Clouds for many hours until they drop from exhaustion and then start to prepare another observing list for tomorrow night!

Auriga Nebulosity

NGC 1931 CL+NB AUR 05 31.4 +34 15

This object is included to prove that every nebula in the sky is not as bright and obvious as the Orion Nebula. This little cluster and nebulosity will certainly do that. Most of the reason NGC 1931 is *not* a showpiece is distance; the best estimates put it at about 5000 light years away. However, lesser objects can prove interesting and a challenge to see some detail. I have enjoyed observing many objects like this that are certainly not bright and gaudy, but are worth some of your observing time.

This image of NGC 1931 was taken by Chris Schur with a 12.5 inch f/5 Newtonian.

With the 13 inch at 100X on a 7/10 night this nebula was pretty bright, pretty large and irregularly round with a bright middle. At low power it appears somewhat like a comet that is just starting to form a tail. Going to higher power really helps. At 220X there are five stars involved and I see this nebula as elongated 1.5X1 in PA 45. The UHC filter does not help the contrast, but averted vision makes it grow in size somewhat.

IC 405 BRTNB AUR 05 16.2 +34 16

This is the Flaming Star Nebula, a glow that surrounds the star AE Aurigae. It seems that this star just immersed itself within the nebula as it moved through the galaxy and has no other reason to be here, except luck. Not all bright stars embedded within a nebula from our point of view are actually part of that nebula's lifeline.

Using the 10 inch f/5.1 Newtonian on a good night, at 60X and no filter there is no nebulosity at this position in the sky. Add the UHC filter and the nebula is seen with direct vision. It is pretty faint, large, and is about one quarter of the field of view with the 22 mm Panoptic eyepiece. It is a somewhat round glow with AE Aurigae on the south side.

This image of IC 405 is by Chris Schur with a 12.5 inch f/5 Newtonian.

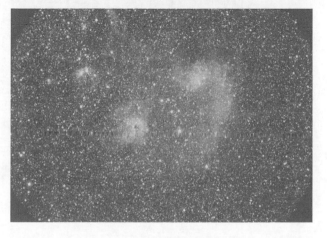

This image of the central part of Auriga was taken by Chris Schur with an 8 inch Schmidt camera. It includes IC 405, IC 410 and obviously lots of stars.

IC 2149 PLNNB AUR 05 56.4 +46 06

Here is another small planetary to test your finding skills. Remember, most planetary nebulae are like IC 2149, a small disk in a Milky Way field. Nobody said it was going to be easy. One of the things that help me in looking for these things is that I see a "planetary sheen." I don't know of another way to describe it, but these little nebulae have a color and texture that is very different from a star of similar magnitude. Once you see it, you will recognize it also. IC 2149 is a great place to start. There is a great piece of trivia associated with this object; it is the only IC object with a description that says "vB" for "very bright."

With the Nexstar 11 on a 7/10 night and the 22 mm Panoptic, this little planetary is just twice the size of the seeing disk at 125X. But it has that "planetary sheen." Moving up to 320X with the 8.8 Ultra Wide Angle eyepiece shows the disk easily and makes this little planetary a light green color. Adding the OIII filter makes the disk about 25 percent larger. The highest power I tried was 420X with the 6.7 mm Ultra Wide Angle eyepiece, which made the planetary disk about four times the size of the Airy disk. At the highest power the light green color is just barely seen. Fortunately, there is a focusing star of about equal brightness just "above" the planetary, a handy coincidence.

While I am at high power, concentrating on this little nebula, a big bright meteor goes right through the field of view and scares me half to death. I have time to look away from the eyepiece and with my naked eye see about 30 degrees of this fireball moving through Orion and Taurus. It flames out and leaves a nice train of smoke that lasts for 3 minutes. Telescopic meteors are a rare and fascinating occurrence. Once my heart rate slowed down enough so I could enjoy it!

This image of IC 2149 is from Chris Schur with a 12.5 inch f/5 Newtonian. The image was cropped and zoomed so that you could see the planetary.

B 34 DRKNB AUR 05 43.5 +32 39

B 34 is an easy dark nebula at 60X in a surplus 38 mm Erfle with gives the 13 inch a 1 degree field of view. The dark area is about half a degree in size and is roundish with several dark lanes winding out of the field to the west. Raising the magnification does not help the view. Even though this dark nebula only got a Barnard rating of 4 out of 6, it stands out from the Milky Way background very nicely.

This image of the central part of Auriga is by Richard Payne with an SBIG-STL11K camera attached to a Canon 200 mm f2.8L lens, 4 hours of exposure. The three Messier open clusters in the image are, left to right, M 37, M 36, M 38. Barnard 34 is the dark oval to the right of M 37. IC 405 is the nebula to the far right.

Cancer Nebulae

Abell 31 PLNNB CNC 08 54.2 +08 55

I chose Abell 31 so that I could discuss objects that are not on the NGC or IC. The main two designations given for this planetary are Abell 31 and PK 219+31.1. George Abell spent many hours looking at the red sensitive plates from the Palomar Sky Survey, taken with the 48 inch Schmidt camera on Mt. Palomar. This survey was completed in the 1950s. Dr. Abell cataloged 86 planetaries on those plates. Lubos Perek and Lubos Kohoutek published a catalog of planetary nebulae in 1965. It has been known as the "PK" catalog ever since. They were able to gather together many observations and data from a variety of sources and make a catalog that proved very useful. The numbering system is based on one degree sections of galactic longitude, not the RA or DEC of that object.

With the Nexstar 11 on a night rated S = 6, T = 7 this planetary is extremely faint, pretty large and has very low surface brightness at 125X. The 10th-magnitude double star that is involved is easy to see and split; it has a hint of nebulosity around it, but not much. Averted vision helps some, but this is still a difficult object to hold steady. The UHC and OIII filters do not help; there must be enough photons to use a filter and these faint nebulae do not respond to the filters.

In the 13 inch on a similar night to the previous observation, it is extremely faint, large, round and not brighter in the middle at 100X. There are three stars involved

This image of Abell 31 was taken by Adam Block/NOAO/AURA/NSF at Kitt Peak.

with the nebulosity; this is using a UHC filter. Realizing that this was going to be a low surface brightness object, I first searched for it at 60X. When it was first seen at this low power with averted vision, I switched to 100X for the observation, but it is tough to see at either power. All that appeared is an irregularly round dull glow that is about 10 arc minutes in size. Ken Reeves says this is a BARF object: that is, Big And Real Faint.

At the Texas Star Party of 2005 I spent a little time with Tracy Knauss and her 18 inch f/4.5 Newtonian. Using a 16 mm eyepiece Abell 31 was extremely faint, pretty large, round and difficult on a night rated S = 5, T = 7, with some clouds moving across the sky.

Canis Major Nebulae

NGC 2359 BRTNB CMA 07 18.5 –13 14

This is another of those nebulae that have been called by several common names. In the Saguaro Astronomy Club we call it the "Duck" Nebula because the outline is the head of a duck. The nebula is also called "Thor's Helmet" because people see that shape within the nebula. The bubble shape that forms the head of the duck is translucent with an orange star in the center. This central star is the Wolf–Rayet star that formed the nebula from its outer layers and then fluoresces the atoms so that the gas glows.

In the 4 inch RFT with a 12 mm eyepiece for 50X and no filter, NGC 2359 is faint, pretty large, irregularly round, and averted vision makes it more prominent. Adding the UHC or OIII filter raises the contrast of the bright area but none of the outer faint nebulosity gets through those filters. The Deep Sky filter provides the best view in the 4 inch refractor. Not only is the bright area displayed with more

This image of NGC 2359 is by Chris Schur with a 12.5 inch f/5 Newtonian.

contrast, but the faint outer sections are now seen. Also, the gentle filtering action of the Deep Sky will allow more of the stars in the view to be observed. This provides a nice frame for the nebula.

Using the Nexstar 11 at 100X with a 27 mm Panoptic eyepiece and no filter shows the nebula as pretty bright and large with a very irregular figure. There are eight stars involved of magnitudes 10 to 12. The bright "duck" figure is obvious and there is some faint nebulosity seen on all sides. Adding the UHC filter makes a lot of difference in this nebula. Three of the faintest stars are not seen with the filter installed, but the observer gains much. The entire field of view is nebulous; there are several bright streamers that go all the way out of the field of view. This is a winter favorite for me and I return to it often.

Sh2-301 BRTNB CMA 07 09.8 –18 29

Stewart Sharpless published a list of 313 H II regions that he found on the Palomar Sky Survey plates. Some are repeats of familiar objects and some are newly

This image of Sharpless 2-301 is by Chris Schur with his 12.5 inch f/5 Newtonian.

discovered gaseous nebulae. This nebula is included as an introduction to the Sharpless catalog. It is one of my favorite "off the beaten path" deep sky objects.

On a good night, S = 7, T = 8, this is an interesting nebula. Using the Nexstar 11 and a 27 mm Panoptic with a 2 inch UHC filter, Sharpless 2-301 is pretty faint, large and somewhat brighter in the middle. Averted vision makes it grow in size. It has a very irregular "three-armed" figure with four fuzzy stars involved within the nebulosity. There is a round area of nebulosity that is connected to three ribbons of nebulosity that end in 10th-magnitude stars with a glow around them.

In the 13 inch it is pretty bright, pretty large and has an irregular shape at 100X on a night I rated 8/10, an excellent night. My first observation of this object was from a somewhat light polluted site and I called it faint. All these observations are with the UHC filter: it helps a lot on this object. This nebula has a three-branch structure with a dozen stars involved. There are a few detached sections of nebulosity that are beyond the 30 arc minute field of view.

Dorado Nebulosity

Well, here we are at the Large Magellanic Cloud (LMC). Other than the Milky Way, it is the brightest member of the Local Group of galaxies. It is a reason to travel to southern skies all by itself. With the naked eye it appears as the cloud that doesn't move! The Tarantula Nebula is a bright spot at one end.

Using my 8X42 binoculars from Jim and Lynne Barclay's backyard in Australia, the LMC is a giant "L" ("L"MC?); the Tarantula is a large bright spot at the short end of the "L." There is a faint spot at the other end of the galaxy. The LMC overall is extremely bright, extremely large, very little brighter in the middle and has a very irregular figure. This giant-sized object is about half of the 6 degree binocular field of view.

There are several entire books that have been written about this fascinating and complex deep sky object. I have neither the space nor the inclination to do such a thing. So I will provide a few of the best observations I made on my most recent Australian trip. All the nights I had to observe rated a seeing and transparency of 7 or 8, so I won't repeat myself during this observations.

This photo of the Large Magellanic Cloud is by Chris Schur with a Schmidt camera.

NGC 1714 LMCDN DOR 04 52.1 –66 56

With the 14 inch f/10 SCT and a 30 mm eyepiece with no filter, this nebula is small but shows a high surface brightness nebula next to a cluster. The nebulosity is elongated 1.8X1 and brighter at one end. The cluster resolves into 40 stars, is pretty large and not compressed. With the 12 mm eyepiece on the bright nebula it is somewhat comet-shaped and the bright end is very prominent. The UHC filter enhances the outer sections of the nebula and makes it larger. This entire nebula is in front of a dark nebula that has a definite edge just to the NW, and then the rich star field starts up again in earnest.

NGC 1769 LMCCN DOR 04 57.7 –66 28

In the 5 inch f/8 refractor with a 30 mm eyepiece, this is definitely a "WOW" field of view; there are three star clusters and much nebulosity in the field. NGC 1761 shows eight stars with direct vision and lots of faint unresolved stars in the background. With averted vision about 20 stars show up in this bright, pretty large, considerably compressed and pretty well detached star cluster. NGC 1763 is a nebula that is bright, large, elongated 2.5X1 and much brighter in the middle. NGC 1769 is another nebula and it is pretty bright, pretty large, irregularly round and much brighter in the middle. There are few places in the sky where a fairly small scope will show this much detail. I have only supplied the data on NGC 1769 above. Once you have that field the rest will fall into place.

Moving up to the 12 inch with the 40 mm eyepiece and no filter, this field is still amazing. NGC 1761 shows 38 stars counted and another 20 appear with averted vision. There are lots of faint and very faint members to this cluster. NGC 1763 is 18 stars involved in a bright, elongated nebula, and there is an obvious double star on the NE side. NGC 1769 is a pretty bright nebula with a difficult double star within it. The double is one bright star with a faint companion that is split with averted vision.

NGC 1874 LMCCN DOR 05 13.2 –69 23

Using the 5 inch with a 30 mm eyepiece and no filter just shows this nebula as a fuzzy area, but going to the 18 mm eyepiece shows this area much more clearly. There are two parts to the nebula, with a delicate curved chain of six stars that connects them. Adding the OIII filter does enhance the nebula, but I like the view without it better. The nebulosity is pretty high surface brightness so it presents better without the nebula filter, in my opinion.

With the 12 inch and a 40 mm eyepiece and no filter, both nebulae are easy with pretty high surface brightness. NGC 1874 is bright, pretty small, elongated NW-SE 1.5X1 and somewhat brighter in the middle. NGC 1876 is bright, pretty small and little elongated N-S. This nebula is in two pieces that are separated by a pretty narrow dark lane. Adding the UHC enhances the two main nebulae and makes them larger and more obvious. It also brings out a faint, pretty small and elongated nebula to the east of NGC 1876. Raising the magnification with the 12 inch f/15

This image of NGC 1874 in the LMC is by Eddie Trimachi.

and the 18 mm eyepiece shows two stars involved within 1874. This would seem to be the "biN" (binary nucleus) from the notes of John Herschel when he was in South Africa.

NGC 2070 LMCCN DOR 05 38.6 –69 06

The LMC is amazingly dense with nebulae and clusters. As I said, this is a very small sampler: just move the 5 inch scope a field or two and there is another entire set of clusters and nebulae to try and identify.

NGC 2070 is a star cluster involved within a nebula and both are outstanding in their way. The star cluster includes several of the brightest stars within reach of any modern telescope. 30 Doradus is the central star in this area and it is several thousand times brighter than our Sun. The nebula surrounding these stars is the largest H II region in the known Universe. NGC 604 in the galaxy M 33 is about half its size, if they were at the same distance.

The Tarantula Nebula (NGC 2070) is amazing – at low power it is surrounded by bright spots and with high power the interior detail inside the Tarantula is overwhelming in its looping strands of nebulosity. It is large enough that as you move the central part of your vision around the field the outer parts of the nebula are seen with averted vision automatically. This effect means that outer detail appears and disappears with your eye's movement – fascinating.

In the 5 inch f/8 refractor with an 18 mm Ultrascopic eyepiece with no filter NGC 2070 is very bright, large, has a bright middle and a very irregular figure. The "Tarantula" shape is obvious and averted vision thickens the hairy legs of the Tarantula shape.

The 12 inch f/15 Cassegrain, a 40 mm eyepiece and no filter provides a great view of this object. The Tarantula shape is 80 percent of the FOV, and 26 stars are involved. Nebulosity is in beautiful curling streamers out from the cluster of stars at the center. Adding the 2 inch UHC filter makes it spectacular. The contrast between the dark lanes and the bright nebulosity is excellent. There are little thin dark lanes all through the Tarantula. The brightest star involved is light orange, that is 30 Doradus.

This image of NGC 2070, the Tarantula Nebula, is by Eddie Trimarchi.

If you are looking for a challenge, this area of the sky can provide a beauty. Using the 5 inch and a 40 mm eyepiece there is an amazing loop of pretty faint nebulosity up and over the Tarantula Nebula that includes three brighter areas within the loop. This loop is about 3 degrees long, the same length as the main bright area of the LMC, but it is thinner and much, much fainter. The UHC does not raise the contrast of this long outer loop: it may be just stars.

Gemini Nebulae

NGC 2392 PLNNB GEM 07 29.2 +20 55

This famous planetary is bright and easy. NGC 2392 is at a distance of 1700 light years and has a diameter of 0.6 light years; the central star is 40X the Sun's luminosity.

People just love to make up names for deep sky objects and this one gets stuck with lots of them. I originally heard of NGC 2392 as the "Eskimo" planetary because the outer layer of nebulosity appears like a furry hood that surrounds a human face. I have never seen the clown face that is supposed to be here, but different observers see a variety of detail within this nebula.

In a small telescope, such as my 6 inch f/6 Maksutov-Newtonian, it is easily seen as non-stellar with a 14 mm eyepiece. The nebula is bright, pretty large and round. It displays a stellar nucleus in the middle of a grey disk. Raising the power on a good night still does not show off any internal detail with this telescope.

A big increase in aperture to Pierre Schwaar's 20 inch Newtonian will really show this planetary at its best. At 350X the amount of detail across the "face" of this object is fascinating. There are several dark markings that surround the central star and there is an obvious gap between the central disk and the "hood" of material that forms an annulus around the outside. Installing a UHC filter makes several of the markings really prominent and enhances the outer "hood" quite a bit.

This image of NGC 2392 is from Chris Schur with a 12.5 inch f/5 Newtonian.

IC 443 SNREM GEM 06 17.8 +22 49

Ah, a supernova remnant. Those really big stars just can't go quietly; they have to blow themselves to pieces and then make the pieces glow in the dark. Lucky for us. I had owned no other filter than a UHC for a decade and then I borrowed Chris Schur's OIII filter and observed this nebula and saw the difference that it made. Even though I generally don't like the effect of the OIII on the stars, it does a great job on some nebulae. Assuming you already have a narrowband nebular filter, make the OIII your next purchase, it is the most useful of the line filters. I will make that obvious with observations of this object.

In the 4 inch there was nothing seen at this location; it is just too faint for the small scope.

Moving up to the Nexstar 11 and a 35 mm Panoptic shows an extremely faint, pretty large, very elongated (3X1) nebula. There are four stars involved. This thick comma-shaped nebula is almost the size of the entire field of view. Adding the UHC helps somewhat, but the nebula is still a low-contrast feature with this filter. Installing the OIII filter makes a lot of difference to the view of this supernova remnant. The contrast is very much enhanced and now it is quite easy to see on a good night.

This image of IC 443 is by Chris Schur with a 12.5 inch f/5 Newtonian.

Abell 21 PLNNB GEM 07 29.0 +13 15

This is another large and faint planetary that George Abell discovered photographically. With the 13 inch at 100X and no filter it is faint, pretty large, elongated and has an irregular shape overall. It is quite noticeable at 100X with the UHC filter. It has a half-moon shape with the southern end brighter and several stars involved. This is the object marked PK205+14.1 on *Uranometria*. It is called the Medusa Nebula in *Sky Cat 2000*. I have a guess why it got that name. Medusa was the snake-haired woman from mythology and a good photograph of this object shows lots of intertwined ribbons of nebulosity. Maybe that is the reason, maybe not.

This image of Abell 21 is from Chris Schur with a 12.5 inch f/5 Newtonian.

Lynx Nebulosity

PK164+31.1 PLNNB LYN 07 57.8 +53 25

PK 164+31.1 is a planetary nebula which has been mistaken for NGC 2474 in several references. *Sky and Telescope* magazine for April 1981 on page 368 tells the story and has a picture of the area. This planetary is very faint, pretty large, not brighter in the middle and has several stars involved at 100X with the UHC filter in the 13 inch scope. The nebula is faint enough that turning on the very dim red flashlight to make a drawing makes the planetary disappear for a few seconds. I had to memorize the field to draw it.

Image of PK164+31.1 by Chris Schur with a 12.5 inch f/5 Newtonian.

Monoceros Nebulae

NGC 2244 CL+NB MON 06 31.9 +04 57

There are several NGC numbers associated with this part of Monoceros, but NGC 2244 is provided because that is the NGC number applied to the cluster within the Rosette Nebula. The Rosette is a large, naked-eye bright spot in the Milky Way and shows lots of detail with some optical aid. It is about 4500 light years distant and 50 light years across. The central hole is 12 light years across.

My 11X80 finder scope will just show a very faint "U" shape of nebulosity around a scattered star cluster. In the 4 inch RFT and a 22 mm Panoptic eyepiece with no filter, the nebulosity is just barely seen; averted vision makes it somewhat more prominent. Also, without a filter, the Rosette nebulosity is never a complete annulus with the small telescope; it is a horseshoe shape around the central star cluster. Adding the UHC filter makes a real difference in the ability to see the Rosette Nebula. Both the filter and averted vision really bring out the big annular shape. I tried the Deep Sky filter and it does not help with this object at all. The Oxygen III filter makes the nebula very prominent by blackening the sky.

This observation is with the 6 inch f/6 Maksutov-Newtonian on a very good night, rated S = 6, T = 8. With the 22 mm eyepiece and without a filter, I counted 28 stars of magnitudes 8 and fainter. The cluster is bright, large, not compressed and includes several bright members in two parallel lines of stars. Using the UHC filter really enhanced the nebulosity. The nebulosity surrounds the cluster on all sides. There are several dark markings on the east side and four thin dark lanes that cut across the nebulosity. A nice view of a famous object.

All the detail seen with the 6 inch could easily be seen in the 13 inch at 60X. On a night I rated 10/10 for transparency I followed several dark "elephant trunks" winding through the nebula. The 2 inch UHC filter makes this object come alive with detail on a sharp night. Without a filter there are 71 stars involved in the nebula.

This image of the Rosette Nebula is by Jon Christensen with a Takahashi refractor.

NGC 2245 BRTNB MON 06 32.7 +10 09

I love coincidences and here is one that I find fascinating. Using the size of telescopes that I have available to me, there are probably a dozen or so objects that are shaped like a comet. I have always enjoyed them. The coincidence is that with the entire sky to spread out these 12 or 15 comet-like objects, two of them are within 2.5 degrees of each other. NGC 2261 is Hubble's Variable Nebula, a famous and interesting object that is probably the best known of comet-shaped deep sky objects. NGC 2245 is nearby and could be the "False Hubble's Variable Nebula," if you are willing to really stretch for an analogy.

With the 4 inch scope and a 15 mm eyepiece this nebula is just barely seen. Going to a 9.5 mm Lanthanum eyepiece will provide a hint of the fan shape. Actually, the nebula blinks on and off with averted vision in the small scope.

With either the Nexstar 11 or the 13 inch it is pretty bright, pretty small and fan-shaped with a star of about ninth magnitude at the apex. The nebulosity is fainter than the real Hubble's Variable, but the star is somewhat brighter and yellowish. Averted vision really enhances the comet shape of this nebula. Neither the UHC nor the OIII filter seems to enhance the nebulosity of this object.

This is an image of NGC 2245, at the bottom, from Jon Christensen with a 12.5 inch R-C telescope and STL11000 camera.

NGC 2261 BRTNB MON 06 39.2 +08 44

OK, so now let's move on to the "real" Hubble's Variable Nebula. Edwin Hubble announced that there were changes in the appearance of this nebula. A study by C.O. Lampland at Lowell (over 1000 photos were taken!) showed that the light and dark edges that were seen in the photos moved at nearly the speed of light. So the changes could not be the movement of dust and gas; the apparent changes are shadows cast within the nebula. As a dark cloud moved around the star R Mon, it cast shadows on the nebula that made it appear to change its shape to Earth-bound astronomers. A recent radio telescope study found a molecular cloud within Hubble's Variable Nebula and that may be the culprit. This nebula is at a distant of 2600 light years and therefore seems to be associated with the Cone Nebula (NGC 2264), which is nearby.

Using the 6 inch with a 14 mm eyepiece will easily provide a view of the wedge shape so familiar in photos from Mt. Palomar. Raising the power with a 6.7 mm shows it as pretty bright, pretty small and with a star at the tip of the triangular or comet shape. At the higher power the comet shape is easy and the western side of the nebula extends further away from the star (R Mon) that is at the tip.

In the Nexstar 11 at 200X it is pretty bright and pretty large. The irregular figure is easy and there is a stellar nucleus at the apex of the comet shape. A dark lane almost cuts completely across the nebula. The western side of the nebula is much brighter, and averted vision makes the entire nebula grow larger.

This image of NGC 2261 is from Carole Westphal/Adam Block/NOAO/AURA/NSF.

NGC 2264 CL+NB MON 06 41.0 +09 54

NGC 2264 is a naked-eye spot in the Milky Way that marks the location of this large, bright and not compressed star cluster. Because of the shape of the brighter stars, this is called the Christmas tree cluster and includes the variable S Mon within the tree shape.

Binoculars or finder will show the tree outline with ease. The cluster is involved in a faint nebulosity that is brightest near S Mon and on the north side of the cluster. In the 13 inch with a 38 mm Erfle eyepiece and the UHC filter the nebula extends for 2 degrees around the star cluster. At 100X with the UHC a dark lane can be seen pushing into the bright part of the nebula. This is the Cone Nebula.

This image of the Cone Nebula is by Jon Christensen with a Takahashi refractor. Hubble's Variable Nebula is at the bottom, centered.

Orion Nebulae

M 42 CL+NB ORI 05 35.3 –05 23

Well, you didn't think that I was going to write a book about nebulae and forget this one, did you? The Orion Nebula is easily one of the most famous objects in the sky beyond the Solar System. It is bright and easy to find, so lots of beginners start their journey with a telescope at this lovely fan-shaped cloud in the Sword of the Hunter. When I am at a public viewing session and I say that this object is 1600 light years away, I try and get that to sink in by saying that the Roman Empire was in its decline when the light left this nebula toward your eye. After providing this profound piece of knowledge a voice in the dark said "Space is like big, dude."

With the 4 inch RFT and a 27 mm Panoptic eyepiece, an observer gets a great view of a busy field. The entire end of the Sword of Orion is visible at one time. The main bright section of the Orion Nebula is easy and there are two nebulae on either site, M 43 to the north end and NGC 1980 at the south. Adding the 2 inch UHC filter makes all these nebulae blend together because of the contrast enhancement from the filter. The field is filled with a nebulous glow that is larger than the 2.5 degree field of view in the RFT refractor. This setup provides an excellent view of a famous part of the deep sky.

In the Nexstar 11 and a 35 mm Panoptic with the big UHC filter, you get a truly "WOW" view. Twenty-eight stars are counted involved within the nebula. The central region is bright and mottled with the dark "fishmouth" feature really standing out nicely. The entire nebula is larger than the field of view, even with the

This image is by Chris Schur with a 6 inch f/3.6 Schmidt-Newtonian and a Canon 10D camera.

35 mm Panoptic. Going to 200X with the 14 mm shows off the mottled effect around the Trapezium stars really well. The "E" and "F" stars in the Trapezium are seen at this power in the Nexstar 11. Installing the OIII filter makes the mottling around the Trapezium even more prominent. It shows up as an interplay of bright nebula and dark grey bands within that glow.

M 78 BRTNB ORI 05 46.8 +00 05

M 78 is one of the brightest reflection nebulae in the sky. There are few enough of these objects that have a field of dust at just the right angle relative to Earth to bounce light toward our telescopes.

This nebula is bright enough to be seen with the 8X50 finder scope on my old 17.5 inch Dobsonian. It is bright, large and fan-shaped at 100X. This nebulosity looks very much like an active comet, with a triangular glow that involves light and dark shading. I'll bet this object has been turned in as a false comet all too often. There is faint nebulosity out of the field of view in several directions. This nebula shows some mottling and has five stars involved. Adding the UHC filter cuts this nebula to about half the size it is without the filter, so there must be almost no emission in the pass bands of the UHC filter.

This image of M 78 is by Chris Schur with a 12.5 inch f/5 Newtonian.

NGC 2024 BRTNB ORI 05 41.7 –01 51

This nebula can be overwhelmed by Zeta Orionis, the "leftmost" star in the Belt of Orion. This is the nebula that is often in a wide-field photo with the Horsehead Nebula, our next object.

On an evening that was rated at 6/10 for both seeing and transparency, NGC 2024 was pretty easy in a 10 inch f/5.1 with a 22 mm eyepiece and no filter. Adding the UHC filter at low power does help this nebula stand out from Zeta Orionis. Much better contrast is seen with the 8.8 mm eyepiece because you can get Zeta out of the field of view. The UHC does *not* help contrast at higher power. The branches of the dark lanes are seen at higher power and averted vision makes a real difference, helping the light/dark contrast. The thickest of the dark lanes comes

This image of NGC 2024 was taken by Chris Schur with a 12.5 inch f/5 Newtonian.

into the nebula on the south side. There are three stars involved within the nebula. Because of the large, parallel dark lanes, Arizona astronomers have taken to calling NGC 2024 the "Tank Track" Nebula.

B 33 DRKNB ORI 05 40.9 –02 28

IC 434 BRTNB ORI 05 41.0 –02 24

I have always said that Barnard 33, the Horsehead Nebula, is a photographic object. It just shows up on a good photograph so much better than an observation with the human eye.

With the Nexstar 11 and a 14 mm eyepiece there is a small notch in the IC 434 nebulous streamer. This nebula is low contrast, so the Horsehead just shows up as a missing section of a very faint glowing band. It can only be seen on nights of excellent contrast. This dark marking has never been easy in any telescope I have owned.

This image of the Horsehead nebula (B 33) is from Chris Schur with a 12.5 inch f/5 Newtonian.

Using Ken Reeve's 20 inch f/5 Newtonian on a night I rated 8 out of 10 for transparency does make some difference. Even without a filter the missing section of the nebula is pretty easy. There is a double star at the "neck" of the Horsehead dark nebula and that points the way to the dark area. It might be easier, but it is still a low-contrast object, even with the big scope and a very good night.

Sh2-276 BRTNB ORI 05 48.0 +01 00

This giant-sized nebula is a low surface brightness object and the arc is centered on the Sword of Orion. It was discovered by E.E. Barnard on photographs he took 100 years ago. It is therefore known as Barnard's Loop. It was created when giant stars in the Sword of Orion spewed out gas and dust by the thousands of billions of tons in the distant past and now you can see that nebula with your naked eye. Yes, I said with your naked eye, and the help of a modern filter. From a very dark location on a great night, just holding the UHC filter up to my eye showed a faint, but not extremely difficult, arc of nebulosity to the east of the Belt of Orion. Make certain that you are well dark-adapted before trying this observation.

In the 8X42 binoculars and no filter, seeing Barnard's Loop is more of a test of the evening than your vision. From a mediocre site on a 5/10 night I could not see any of the nebula. On a much better 8/10 night it was not difficult and about 4 degrees of the arc were seen with averted vision in the binoculars.

The Sharpless 2-276 designation seems to cover the entire arc of nebulosity. In the brand new 35 mm Panoptic that I bought just for this type of observation, it is very faint, extremely large and looks like a very, very elongated streamer with the 13 inch. A little cluster, NGC 2112, is a good marker for the nebulosity; it passes right over this clump of a dozen pretty faint stars. Then the streamer of nebulosity extends about 3 degrees on either side of this cluster. The UHC filter helps very much, and so does averted vision: this nebulous streamer shows much better contrast out of the corner of my eye.

This image of Barnard's Loop was taken by Richard Payne with a Pentax 6X7 camera and a 165 mm f/2.8 lens.

Puppis Nebulosity

NGC 2438 PLNNB PUP 07 41.8 −14 44

M 46 and NGC 2438 offer an opportunity to understand that we see the Universe in two dimensions, even though it is at least a three-dimensional reality. The planetary nebula NGC 2438 is at almost half the distance of the star cluster M 46. I know that the nebula looks like it is right on the NE boundary of the open cluster. It is an optical illusion. The planetary is about 2500 light years away and the cluster is over 4000 light years distant.

The actual star that illuminates the nebula is 18th magnitude, but there is an 11th-magnitude star in the middle of this nebula. This is also just a chance alignment.

Using the 10 inch f/5.1 with an 8.8 mm eyepiece the nebula is bright, large and elongated 1.2X1 in a PA of 75 degrees. The bright middle star of this planetary is held steady about 50 percent of the time. The nebulosity is light green in color. Going to the 4.7 mm eyepiece is a nice view of the annular nature of this planetary, even though the ring is not complete. The annulus is brighter on the south side.

This image of M 46 and NGC 2438 is by Chris Schur with a 12.5 inch f/5 Newtonian.

Moving way up in aperture to a 25 inch f/5 with a 12 mm eyepiece for 300X shows lots of detail in this little planetary. It is light green and three stars are seen within the nebula. Two of these stars are obvious; one star is more difficult. The glow of the nebula is annular and the central section is light grey. There are two layers of nebulosity that surround the central star.

NGC 2467 CL+NB PUP 07 52.5 −26 26

The star cluster is very nice and would generate observers if it where alone, but there is some bright nebulosity associated with this cluster. The notes on this cluster say that it is within the Puppis OB1 association. Sixty years ago Bart Bok and several others were working on the problem of where the spiral arms are located within the Milky Way. They found many of these groupings of bright, blue-white stars that they labeled "OB associations." They used them to trace out the spiral arms in our Galaxy. What they mean to an observer is that one of these associations is going to contain a concentration of bright, blue-white stars in that area of the sky.

Using the 4 inch with a 12 mm eyepiece and the Deep Sky filter makes this an easy-to-see round nebula. It is pretty faint, pretty small, round and has three stars involved. Averted vision makes it grow. Without the filter it is just barely visible in the small scope.

NGC 2467 in the 13 inch is a bright, pretty large and pretty rich cluster. I counted 31 stars at 100X on a 7/10 night. This cluster was easy in the 11X80 finder. The nebula was seen without the UHC filter to start, but adding the filter made the nebula much better. There is a bright, round spot of nebulosity on the southwest side of the cluster and several pretty bright streaks on the northeast sections. Covering my head with a dark cloth and using the UHC filter allowed me to see that the entire field of view was nebulous to some degree. To top it off, there are several dark lanes winding their way through this region. Take a look at this little-known cluster and nebula.

This image of NGC 2467 was taken by Chris Schur with a 12.5 inch f/5 Newtonian.

Pyxis Nebulosity

NGC 2818 OPNCL PYX 09 16.0 –36 38

NGC 2818A PLNNB PYX 09 16.0 –36 36

The NGC number 2818 was associated with both this open cluster and the planetary nebula at the edge of it. Someone added the suffix "A" to the nebula so that they would have separate numbers. Regardless of human record keeping, the two objects are associated in the sky and are about 10,000 light years away. This is a very nice rich area of the Milky Way and this cluster and planetary stand out pretty well. They are certainly worth chasing on a sharp night.

In the 13 inch scope, this cluster is pretty faint, pretty large, somewhat elongated and 16 stars were counted across a hazy background of very faint stars at 135X on a 5/10 night. On a much better night (7/10) this cluster was seen as pretty bright and I could resolve 34 members at 150X. The planetary nebula is pretty bright, pretty large, a little brighter in the middle and has fuzzy edges at 150X. It is located on the eastern edge of the open cluster. A few dark lanes are seen in the planetary at 285X; it is light green in color.

From Australia, this object really shines (sorry about that). In the 5 inch refractor with an 18 mm Ultrascopic eyepiece and no filter, the cluster resolves into 10 stars of magnitudes 10 and fainter. It is pretty faint, pretty large and has two double stars involved. The nebula is pretty large, faint, round, shows no central star and averted vision makes it larger. The OIII filter makes a big difference and now the planetary really stands out and it easy to see, pretty large and round.

Moving up to the 12 inch Cassegrain with no filter gives a great view of the cluster; 42 stars are resolved and it is pretty bright, pretty large, pretty rich, compressed and shows many pairs within the cluster with a 40 mm eyepiece. Going to the 30 mm eyepiece the planetary is obvious, pretty bright, pretty large, a little elongated 1.2X1 and a hint of a dumbbell shape is seen. The UHC filter makes the contrast much better. The central star of the planetary was never seen.

This image of NGC 2818 is by Jim Barclay with a 14 inch SCT.

Taurus Nebulosity

M 1 SNREM TAU 05 34.5 +22 01

The Crab Nebula is one of the most studied objects in the sky. Once scientists realized that this little glowing cloud was the remains of a star that exploded in the year 1054, they really went all out to explain the inner workings and chemical makeup of this nebula. And the Crab did not disappoint them. There are a wide variety of atoms and molecules within the nebula. A fascinating neutron star beams out huge amounts of energy from the center of this growing cloud.

With the 4 inch scope and a 22 mm eyepiece it is pretty bright, pretty small, elongated 2X1 and shows off a pretty high surface brightness. Moving up to the Nexstar 11 and a 35 mm eyepiece it is bright, pretty large and much elongated 2.5X1. There are no stars involved within the Crab Nebula at low magnification. Raising the power with a 14 mm Ultra Wide eyepiece shows off a fascinating grey-green color. There are four stars involved, all at the edges; none of these stars is the central pulsar, which is 16th magnitude. High power does show a hint of the brighter "crab-like appendages" that gave this object its nickname. The most prominent of these is a thin S-shape that runs from one end of the nebula to the other. These brightenings within the Crab Nebula are low surface brightness details that are helped by averted vision.

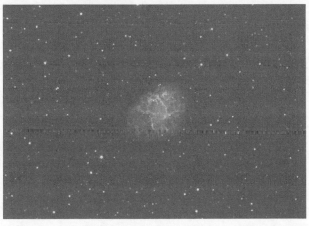

This image of the Crab Nebula is by Chris Schur with a 12.5 inch f/5 Newtonian. It is taken in the wavelength of Hydrogen Alpha emission.

M 45 CL+NB TAU 03 47.0 +24 07

I have done many public viewing sessions over the years and there are lots of people who think that the Pleiades are the Little Dipper. And I can see how that could happen. It does indeed look like a small dipper with the bowl up and the handle down as it rises in autumn and early winter for those of us in the northern hemisphere.

The most obvious reflection nebulae in the sky are associated with the stars in the Pleiades cluster, Messier 45. These dust clouds are left over from the formation of the stars in the cluster and as that starlight reflects off the dust it creates a glow around the stars.

From a dark site almost any optical aid will show a glow around Merope; in my 8X42 binoculars the view is terrific. I counted 40 stars involved in the cluster and the nebulosity around Merope was quite easily seen.

This is one of the objects I was thinking about when I acquired the 100 mm refractor. It takes a wide-field telescope to view all of the Pleiades at one time. The good news is that it did not disappoint me. The little scope provides a great view of a great cluster with nebulosity. The entire Pleiades cluster is displayed and there is some room around it. Most of the major stars have some glow around them, Merope being the most prominent. The stars that are the "handle" display no nebulosity. Averted vision and the 22 mm Panoptic eyepiece really enhance the glow around the stars.

This image of the Merope Nebula is by Chris Schur with a 12.5 inch f/5 Newtonian.

Vela Nebulosity

NGC 2736 BRTNB VEL 09 00.4 –45 54

Gum 12 SNREM VEL 08 30.0 –45 00

About 12,000 years ago there was a very bright star in the constellation of Vela. This huge star blew itself to pieces in a supernova explosion, Type II. Today the core is a pulsar that cannot be seen with any visual telescope. However, the outer portions of the star saturate this part of the sky with a fascinating nebula that is truly huge. Even though it is only about 800 light years distant, it has had all those centuries to spread out and so it is a low surface brightness and very diffuse nebula. NGC 2736 is the brightest portion and it is known by the nickname "Herschel's

Pencil" from the description given by John Herschel while observing in South Africa, early in the nineteenth century.

With the 5 inch f/8 refractor in Australia and a 40 mm eyepiece with a UHC filter there is lots of faint nebulosity, three fields of view worth. It blocks off the dense star clouds in this region, so it is not a bright shining nebula as much as streams of star-poor regions with a very faint glow. Installing the 30 mm Ultrascopic eyepiece and an OIII filter makes this a fascinating field of view. NGC 2736 is pretty bright, large, extremely elongated and easy with the filter. Herschel's Pencil is just seen without the filter. Averted vision helps, but the OIII filter provides the best view without doubt.

Now over to the 12 inch Cassegrain and a 40 mm eyepiece with a 2 inch UHC filter. Herschel's Pencil crosses the entire field. It is pretty bright, very large and extremely elongated in a PA of 15 degrees. Averted vision helps contrast; this is a low surface brightness object. The east side is a very sharp edge to the nebulosity; it must be a dark lane stopping the glow at the place where the bright nebula and the dark nebula meet. There are five stars involved in the nebula, two of them pretty bright. Even with this wide-field eyepiece the nebulosity of this region never stops! I have moved the scope five fields of view is all directions from NGC 2763 and there is pretty bright to very faint nebulosity in all directions. The entire region is delightful to observe.

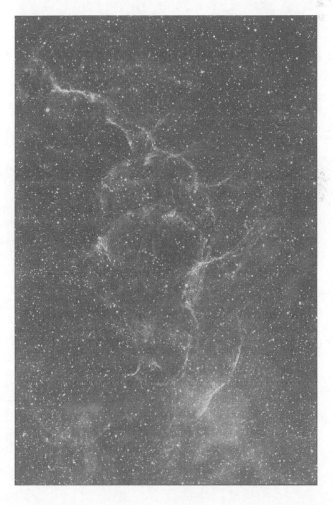

This image of the Vela Supernova Remnant is by Jim Barclay with a 180 mm lens at f/2.5. NGC 2736 is the brightest streak of nebulosity near the bottom of the photo.

This close-up image of Herschel's Pencil was taken by Eddie Trimarchi.

NGC 3132 PLNNB VEL 10 07.0 –40 26

From the complex structure on photos NGC 3132 has been called the Eight-Burst Nebula. It was found that the central star, HD 87892, magnitude 10, A0, is not truly the illuminating star of the nebula; the UV radiation is supplied by a 15.8-magnitude dwarf companion 1.6 are seconds distant from the bright central star. This is one of the brightest planetary nebulae. It is conspicuous at eighth magnitude and comparable in size to the Ring Nebula.

From Arizona in the 13 inch it is bright, large, elongated 1.5X1 in PA 15 and much brighter in the middle with a stellar nucleus at 150X. This is a very nice planetary with a 10th-magnitude central star that is obvious at all powers. Averted vision makes the nebulosity grow around the star. I have seen this nebula as either grey or light green on every occasion I have observed it.

From Australia it is easy in the 5 inch refractor. With an 18 mm eyepiece the disk is elongated and the central star is obvious. Averted vision makes the nebula larger.

In the 12 inch with a 30 mm eyepiece and no filter it is bright, pretty large, elongated 1.5X1 in a PA of 15 degrees and the 10th-magnitude star is obvious. There is a hint of annular structure that is "filled in" with nebulosity that is fainter than the elongated ring around it. Going to higher power with the 18 mm eyepiece really provides a WOW view. The annular structure is easy and there is now a very faint star at the edge. Inserting the OIII filter shows a little bit more nebulosity outside the annulus and it suppresses the star so the detail in the nebula is easier to see. With no filter this planetary is a beautiful aqua color.

This image of NGC 3132 is by Daniel Verschatse.

Chapter 10

Northern Spring (Southern Autumn) Nebulae

In the northern hemisphere springtime is Galaxy time. This is very much the shortest chapter among the nebula observations because Virgo, Leo, Coma Berenices and Ursa Major dominate the northern sky. However, the folks in the southern hemisphere are treated to a very different "view." During April and May they are observing in Centaurus, Crux and Carina. Lots of Milky Way goodies if your latitude has a negative sign.

Carina Nebulosity

NGC 3324 OPNCL CAR 10 37.3 –58 40

This is a good time for a quick discussion of data sources. As I write this, it is obvious that the SAC database version 7.2 has an incorrect "type" for this object. It says open cluster and this cluster includes an obvious nebula. That will be corrected in the next release. No data source is infallible.

Obviously, all these observations are from Australia. With the 5 inch and a 40 mm eyepiece with UHC filter, the nebula around NGC 3324 is obvious. The cluster is not prominent, but the nebula really stands out. In the 30 mm Ultrascopic eyepiece with no filter the nebula is pretty large, pretty faint, round and shows 16 stars involved. Higher power with the 18 mm shows a two-lobed feature. There is a double star on the north side that is involved within a pretty bright and elongated area of nebulosity. There is also a bright star on the south side that is surrounded by a round area of nebulosity. A beautiful deep orange and blue double star is located between this nebula and Eta Carinae.

Moving to the 12 inch with a 40 mm eyepiece and no filter provides a good view of this excellent nebula. It is 80 percent of the field of view, pretty bright, large and elongated 1.5X1 N–S. There are 20 stars involved in the glow and a dark lane cuts off the west side. What a field! And Eta Carinae is right next door – Wow.

This image of NGC 3324 is by Jim Barclay with a 14 inch SCT.

NGC 3372 BRTNB CAR 10 45.1 –59 52

When it gets dark and the Eta Carinae Nebula, the Coal Sack and the Magellanic Clouds are up, you are not in Kansas anymore. The magnificent Eta Carinae Nebula is one of the very best deep sky objects to be observed from Earth and those south of the equator have the best views, no doubt about it. This magnificent nebula is one of the main reasons that I traveled to Australia twice and it has never failed to impress me.

Let me start with the naked-eye impressions and then work my way to binoculars and telescopic views. On a clear night in Jim and Lynne Barclay's backyard, about 90 miles from Brisbane, the Eta Carinae Nebula is easily naked-eye. There are three clusters that surround it: IC 2602, NGC 3114, and NGC 3532. All are excellent star clusters, each has a different character and they are worth examination. But this is a book about nebulae.

With my Orion Savannah 8X42 binoculars, NGC 3372 is a nebula that is about twice the size of the Orion Nebula and brighter because of the 16 stars involved. Being in Australia in April, I could just swing the handy little binoculars from one to the other for comparison. For binoculars, Eta Carinae wins overall for brightness and detail. The star Eta Carinae is the brightest in the nebula glow and is obviously light orange in binoculars! There is an easily seen dark notch within the nebulosity, which cuts the nebula into one third, two thirds portions. To the east of the nebula is an orange star and to the north of it is a beautiful very wide orange and blue pair. This is a fascinating and rich view of a famous object.

The star cloud in which Eta Carinae is embedded is *rich*, an amazing profusion of faint and very faint stars in the 8X42s. The star cloud is elongated 5X1 and cut off at top and bottom by dark lanes that are easy in the little binoculars. It is much like the Cygnus Star Cloud.

The huge dark lacy nebula that surrounds Eta Carinae is fascinating in binoculars. It loops up and over two bright clusters on either side: NGC 3532 on the "left" and up and over NGC 3114 on the "right." There are many delicate patterns of darkness, like a lace table cloth hiding the stars beyond.

Using the 5 inch f/8 refractor and a 30 mm eyepiece with no filter is most certainly a "WOW!" view; there is no other place like this in the sky. Nebulosity covers the entire field of view; it fades out to the south into a very rich field of stars. About 100 stars are involved within the nebula, delicate pairs and chains throughout. Dark markings cut the nebula into one third, two thirds pieces and there are lots of "elephant trunks" of dark nebulosity. There are 28 stars in the triangular "boomerang" of nebulosity that includes Eta Carinae – the star. There are 11 stars resolved in Bochum 10, a pretty compressed cluster off to the west. There are two matched pairs within the cluster. Sliding in the 2 inch 40 mm eyepiece and the same-sized UHC filter frames the nebula perfectly. With loops of light and dark nebulosity intertwined in a very rich star field, this view is truly amazing. Once you have spent some time with the Eta Carinae Nebula you will never mistake it for any other location in the sky.

In the 12 inch Cassegrain with the 40 mm eyepiece, the bright "boomerang" nebulosity is 40 percent of the field of view. There is a tiny nebula that surrounds Eta Carinae; it was given the name Humoculous. There are 32 stars in this area that includes the Humoculous. The Humoculous is about three times the size of the seeing disk and is a beautiful yellow-orange color. The cluster Trumpler 16 has 24

This photo of the Eta Carinae region was shot by me, with a lot of help from Jim Barclay; it is a 20 minute exposure with a 200 mm lens.

This closeup of the area around the Keyhole Nebula is by Eddie Trimarchi. The Keyhole is the dark area that is centered, the Humoculous is the bright nebula above it.

stars and is pretty bright, pretty large, compressed, pretty rich and is located just at the end of the nebulous boomerang shape. On a steady night with the 12 mm eyepiece in the 12 inch, I saw detail within the Humoculous, the star Eta Carinae was a small disk at the center of two opposing bulging disks of light orange nebulosity. Also, there were small, thin dark lines involved in that nebulosity, like lines of latitude on a globe.

Oh yes, the "Keyhole"; and you thought I forgot. It is a dark feature just to the west and north of the star Eta Carinae. It is indeed dark; only, say, six pretty faint stars within the dark area. The Keyhole is about 15 arc minutes long and the southern 5 arc minutes or so is the darkest part of this famous dark nebula. On a good night I found it unmistakable even in the small 5 inch telescope. It does indeed have the shape of an old keyhole, or an upside-down and thick exclamation point.

In the afternoon before the last night that I was in Australia, Jim got a package from the US. It was his Stellarcam video camera and he was thrilled. We removed the secondary mirror and hooked it to the Celestron 14 inch with a Starizona Hyperstar lens attachment. After a little tweaking to get it up and working the views were terrific. Using the Stellarcam is watching TV – but the show is the Universe. Omega Centauri absolutely fills up the field of view with tiny stars. Eta Carinae shows off the Keyhole dark feature easily and I counted 75 stars in the bright "boomerang" of nebulosity.

Centaurus Nebulosity

NGC 3918 PLNNB CEN 11 50.3 –57 11

Color at the eyepiece of a telescope is very subjective. Observers from years past included lots of color in their descriptions of double stars. I have seen some color in doubles, but let's not get into that. There is also some color that I see in planetary nebulae. If the spectrum includes lots of OIII emission then the nebula is light green to aqua to my eye. This is one of the very few nebulae that I have ever seen as blue. If you don't live at a southern latitude you will need to travel to look for yourself.

In the 5 inch with an 18 mm eyepiece the disk is just seen. It is bright, small, round and contains no star. At higher power with the 12 mm eyepiece the disk is much easier and it is a light blue-green color.

With the 14 inch SCT and the 18 mm eyepiece this is a very nice light blue disk in a pretty rich field of view. The central star was never seen at any power. The disk is little elongated at 1.2X1 and averted vision makes it grow and show a faint outer glow. The outer nebulosity is much easier with averted vision. On a good night this is a beautiful Easter Egg in a rich field of view.

This image of NGC 3918 was taken by Geoff Johnston with a Vixen 4 inch apo refractor. It is a 30 second exposure with a Canon 300D camera at prime focus.

Corvus Nebulosity

NGC 4361 PLNNB CRV 12 24.5 –18 48

Using the 13 inch scope, NGC 4361 is bright, large and somewhat elongated (1.5X1) in PA 90. The central star is obvious at all powers. Going to 220X brings out an almost "mottled" effect across the face of this planetary nebula, a strange effect for this type of object. Most planetaries appear smooth at high powers; this one does not. It is grey in color and the UHC filter helps the contrast somewhat.

At the 1996 Texas Star Party I got to spend some time with Tom Clark's 36 inch f/5 "Yard Scope." With a 27 mm Panoptic eyepiece this planetary is bright, pretty large and somewhat elongated. The disk shows some mottling at low power, for a scope with 180 inches (4572 mm) of focal length! Using a 14 mm eyepiece shows this planetary as grey-green in color, bright and large. The central star is in a void or dark area that is "football"-shaped. This elongated dark region around the central star is shown in the image.

This image of NGC 4361 is by Elliot Gellman and Duke Creighton/Adam Block/ NOAO/AURA/NSF.

Crux Nebulosity

Coal Sack DRKNB CRU 12 53.0 –63 00

Certainly the best-named of the dark nebulae, the Coal Sack is next to the South-ern Cross and makes for one of the most striking naked-eye sights anywhere in the sky. Its estimated distance is 550 light years, which makes it about 60 light years wide. If all those values are a good approximation then this is the closest dark nebula to the Earth.

My most memorable observation of the Coal Sack is from the deck of the cruise ship *Dawn Princess* off the coast of South America, waiting for a solar eclipse. The Coal Sack is easily seen naked-eye, just "beside" the Southern Cross. I believe that the Bowl of the Pipe Nebula in Ophiuchus is darker and more contrasty. However, this famous dark marking in the Milky Way does stand out against a dense and starry region. A pair of 10X50 binoculars helps the view quite a bit. The Coal Sack is a dark oval in the Crux Milky Way. The ship does move about a little as it is makes its way toward Aruba and so I have widened my stance on the foredeck and that seems to allow me to hold the binoculars steadier.

On dry land, from Australia, in the 8X42 binoculars I can see 26 stars involved within the Coal Sack. The dark nebula is not one solid piece; it is striations of thick dark lanes with stars involved. It just fills the 8X42's field of view. The Jewel Box cluster is just above the Coal Sack and it shows six stars and is bright and compact.

The aboriginal tribes in Australia created several "dark animals" within the Milky Way. Because their ancestors had very few street lights or football stadiums, they could see the carbon clouds in our Galaxy with ease. So, in the same way that Greek and Roman astronomers created their mythology in the sky, the tribesmen of Australia did the same. The Dark Emu is easy once seen, an unmistakable and huge dark bird that goes from the Coal Sack to the tail of Scorpius . . . one of the sky's most fascinating views with no optical aid. Using the 8X42s along the massive dark bird is fascinating, a view of light and dark, sparkling star clouds and nebulous glows along the way. There is an eye within the Coal Sack and the "beak" extends to the bottom of the Southern Cross, now down the neck that includes Alpha and Beta Centauri, the body thickens and the legs move off toward southern Scorpius, the tail feathers above that toward the head of the Scorpion. Absolutely fascinating.

This photo of the Dark Emu was taken by me; it is an 8 minute exposure with a 28 mm lens. The Coal Sack is at the top of the frame and Antares and the False Comet region (NGC 6231) is at the bottom of the frame. Alpha and Beta Centauri are in the "neck" of the Dark Emu.

A nearby star has the variable star designation of DY CRU. And, while I am in the area, I must mention Ruby Crucis. This carbon star was given its nickname over 100 years ago and the variable star name more recently. It is a carbon star located 2 arc minutes west of Beta Crucis. The B-V magnitude of this star is an amazing 5.8, making it one of the reddest stars in the sky. In the 12 inch with a 40 mm eyepiece it is a "Wowwie Wow" view: a gorgeous dark orange star in the field of a second-magnitude blue-white star. There is no other place like this that I know of in the sky. If you are a fan of red carbon stars, don't miss this one.

I am still raving about these dark lanes. A fascinating 15 degree long chain of darkness winds its way from the Coal Sack to envelop IC 2602 (Southern Pleiades). It is just seen naked-eye but really comes to life in the binoculars. This is with a 60 percent illuminated Moon, just rising over the eastern horizon. Yes, things rise in the east when you are in Australia.

This photo of the Coal Sack was shot by me; it is a 12 minute exposure with a 135 mm lens.

Hydra Nebulosity

NGC 3242 PLNNB HYA 10 24.8 –18 38

Many years ago this planetary got the nickname "Ghost of Jupiter" because it is somewhat oblate in shape and about the same angular size as the giant planet. Beyond the Messier objects, this is one of the brightest deep sky objects and virtually any small telescope can see it on a clear night.

With the Nexstar 11 at 200X, NGC 3242 is bright, pretty small and somewhat elongated (1.2X1). Moving up to 320X shows an excellent view of this famous object. The effect of an eye looking back at you is easy to see. The disk of this beautiful planetary is a light aqua color and the central star is seen about 20 percent of the time.

In 1979, when I was first starting to observe the sky with any sense of knowledge about what I was seeing, I had the good fortune to spend some time with Richard and Helen Lines. They discovered Comet Seki-Lines in 1965 and established an observatory in the small town of Mayer, Arizona. The telescope installed in that observatory was a 16 inch f/8 Newtonian that provided excellent images. One night we put NGC 3242 in that scope and the results were stunning. At 160X with a 20 mm eyepiece the "CBS eye" effect is easy across a disk that is electric blue-green in color. Going to a 12 mm eyepiece gives 275X and makes the color a little duller, but the interior detail is much more obvious at higher power. Averted vision shows a faint outer shell that makes this planetary almost double in size with the bright inner regions surrounded by an elongated and faint outer glow.

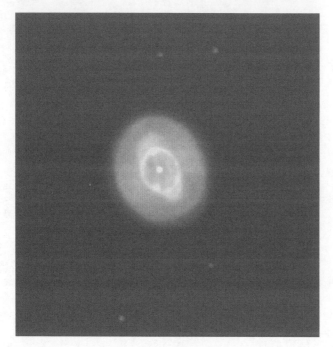

This image of NGC 3242 is by Adam Block/NOAO/AURA/NSF.

Musca Nebulosity

NGC 5189 PLNNB MUS 13 33.5 –65 58

This planetary nebula is about 3000 light years distant. Another scientific finding indicates that the central star is a binary. That will explain the bizarre shape from the two stars interacting with the gases that are spraying out from the middle of NGC 5189. Several previous observers have mentioned that it looks like a barred spiral galaxy and therefore this is called the "spiral planetary."

With the 14 inch f/10 SCT and an 18 mm eyepiece with no filter it is bright, pretty large, and shows a very irregular figure with five stars involved. The "barred spiral" shape is easy at this power. There are dark markings in the nebula with direct vision. A tiny dark lane cuts through the arm on one side. Averted vision makes it much larger; there is an obvious faint outer nebulosity. Adding the OIII filter enhances the contrast of the brightest part of the nebula and makes the outer sections much more obvious now. An excellent view of a unique object.

This image of NGC 5189 is by Eddie Trimachi.

Sandqvist 149 DRKNB MUS 12 25 –72 00

The nickname of this object, the Dark Doodad, seems rather silly and I agree with some observers that the naming of deep sky objects has gotten a little out of control. I choose Sandqvist 149 as the professional designation of this object because I think that people who discover these objects ought to be remembered, but you can't hold back the sea with a broom and lots of people know this dark marking as the Dark Doodad.

This photo of Musca was taken by me; it is a 10 minute exposure with a 135 mm lens.

This dark lane is easy in the 8X42 binoculars; it is immediately obvious. The dark lane is L-shaped (dare I say boomerang-shaped?) and is right in the middle of the Musca quadrilateral (a four-sided figure with all sides different lengths). It goes from just north of Gamma Musca and NGC 4372 and loops over to just above NGC 4833. Both NGC objects are globulars. The western side of this dark nebula is longer and somewhat easier to make out. Going to the 11X80 binoculars shows the dark lane much easier, as well as the globular clusters at either end. Averted vision in the big binoculars really brings out the dark chevron shape.

From Maidenwell Observatory in the 14 inch SCT it is easy to see with a 40 mm eyepiece, but it is eight fields of view long! It appears as a fascinating dark river that winds its way through rich star fields.

Ursa Major Nebulae

M 97 PLNNB UMA 11 14.8 +55 01

Of the four planetary nebulae in the Messier catalog (can you name them all?), M 97 (the Owl Nebula) is the faintest. However, it does have some detail within the nebula and this makes it interesting. So, while the owner of a small telescope can be happy to find and observe M 97, larger telescope users will look for the "owl eyes," two dark regions within the glow that give this nebula its popular name.

Lord Rosse, the Irish gentleman who built the largest telescope in the world in the middle of the 19th century, was first to see the owl eyes. There is a mystery associated with his observations. He mentions a star within each eye, but one of the stars has disappeared since his observations. I plan to check this out when I get an evening on a 72 inch telescope!

With the 4 inch RFT M 97 is pretty faint, small and round at 50X on a great night. The disk is smooth and I do not see the dark owl eyes or any stars involved.

The image of M 97 is by Chris Schur with a 12.5 inch f/5 Newtonian.

Using the largest telescope I ever built, an 18 inch f/6, the Owl Nebula is bright, pretty large, round and somewhat brighter in the middle at 100X without the UHC filter. Using the filter at 165X the "owl eyes" come and go with the seeing, and there is a bright section between the eyes that includes a central star. Both the UHC and the OIII filter make the Owl Nebula quite a bit larger in size. I would estimate that the filters make M 97 about 50 percent larger. The added nebulosity is fainter than the main features, but it can be held steady with direct vision.

Extragalactic Nebulae in M 101

All of the nebulae detailed in this book so far have been within the Milky Way Galaxy and relatively close, on the biggest cosmic scale. Let's move much farther away to the spiral galaxy M 101 in Ursa Major.

All of these diffuse nebulae are located in the spiral arms of M 101. Most were discovered by Lord Rosse with his giant telescope in Ireland. It takes some aperture to chase H II regions in other galaxies, but it certainly is an interesting view when you realize what you are observing: a giant-sized Orion Nebula or Eta Carinae that is millions of light years distant. We do live in a fascinating Universe.

The four nebulae in the arms of M 101 that I viewed came in two pairs nearby to each other. Therefore, I will discuss them in that format.

NGC 5447 GX+DN UMA 14 02.5 +54 17

NGC 5450 GX+DN UMA 14 02.5 +54 16

NGC 5447 and 5450 are very close together. On a night rated 7/10 for seeing and 9/10 for transparency, the two nebulae are seen as pretty faint, pretty small and much elongated (2.5X1). Averted vision makes them much more easily seen. That observation is with my Nexstar 11 at 200X. Using A.J. Crayon's 14.5 incher with an 8.8 mm UWA eyepiece shows the two nebulae much more easily and they are separated by a thin dark lane that is seen about 10 percent of the time. With Bob Kepple's 22 inch Newtonian at the Texas Star Party, the two nebulae were pretty bright, pretty small and elongated 3X1. In the larger scope the dark lane was held steady at 200X.

NGC 5461 GX+DN UMA 14 03.7 +54 19

NGC 5462 GX+DN UMA 14 03.9 +54 22

With the Nexstar 11 with a 7/10 night, both of these objects can be seen at 200X. However, NGC 5461 is difficult and was never held steady. NGC 5462 was seen as the brightest spot in the spiral arms of M 101 on this night.

With A.J.'s 14.5 inch and a 8.8 mm eyepiece, NGC 5461 has a lower surface brightness and appears to be more involved in the spiral arm. NGC 5462 is observed as pretty bright, pretty small, elongated 2.5X1 and has a brighter middle. NGC 5462 is somewhat mottled and averted vision makes it grow.

This image of M 101 is from Chris Schur with a 12.5 inch f/5 Newtonian. NGC 5461 and 5462 are at the top left edge of the image. Point your Internet browser at www.ngcic.org and go to the public NGC database. Search for any of these NGC numbers and you will find that Bob Erdmann has prepared an excellent finder chart with labeled NGC numbers for these clusters and nebulae in the spiral arms of M 101.

Chapter 11

Northern Summer (Southern Winter) Nebulae

Ah, saving the best part for last. When it is summer in the north, the center of the Milky Way is as far above the horizon as it is going to get and now is the time to take on Scorpius and Sagittarius, both with lots of nebulous gas and dust clouds to enjoy. Don't overlook the smaller constellations: Corona Australis and Vulpecula, as an example, have several excellent nebulae that are worth some of your time under dark skies.

Aquila Nebulae

NGC 6751 PLNNB AQL 19 05.9 –06 00

Even though I have viewed this object several times over the years, I have never chased it with a small or medium aperture. So I will borrow an observation from A.J. Crayon, my observing buddy for over 25 years. His 8 inch (200 mm) f/6 has seen many hours of soaking up starlight. Concerning NGC 6751, A.J.'s notes say it is pretty faint at 12th magnitude, very small at 15X13 are seconds. This little nebula lies in a magnificent Milky Way field.

My observing notes for NGC 6751 begin with my Celestron Nexstar 11 on a good, but not great, night. I rated the evening at 6 out of 10. It is seen as a small dot with a 22 mm Panoptic eyepiece. Moving up to 320X with the 8.8 mm UWA it is pretty bright, still pretty small and little elongated 1.2X1. There is a stellar core at high power and that star is seen 20 percent of the time. This little nebula appears as a light green disk afloat in a pretty rich field of stars. It was never seen as annular. The central star magnitude of this object is 13.3, so it was good to know that my new scope picked it out.

In the 13 inch Newtonian it was pretty large, bright, somewhat elongated and greenish at 135X. This is a very nice planetary that will show off its central star at 270X. Covering my head with the monk's hood shows the central star as pretty faint, but it can be held steady. There is a star of similar magnitude on the eastern edge of the nebula, seen at 220X. This is on a night I rated 8/10 for seeing and 9/10 for transparency in the central Arizona Mountains. On that exceptional evening, this object appeared to me as one of the best objects in Aquila.

A very nice gentleman named Jim Christensen let me observe with him and his 25 inch f/5 at the Texas Star Party in 2005. I rated the night at S = 7 and T = 8, a

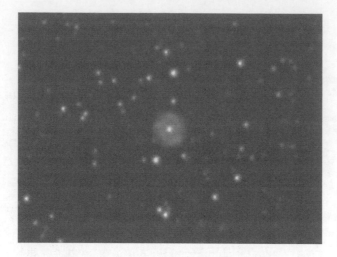

This image of NGC 6751 was taken with a 20 inch R-C telescope by Adam Block/NOAO/AURA/NSF.

good night. With an 11mm eyepiece this planetary was bright, pretty large and round. The central star is seen 100 percent of the time. There is a faint outer shell of nebulosity that is averted vision only. Going back and forth between averted and direct vision makes the outer faint glow come and go. Adding an OIII filter shows the disk as larger, so some of the outer nebula is held steady with the filter in place. The color of this disk is light green, without the filter.

NGC 6804 PLNNB AQL 19 31.6 +09 14

With the 13 inch at 135X this planetary is bright, pretty large and somewhat comet-shaped. Using 200X shows one star at the tip of the comet shape and another dimmer star involved at the edge. This is a nice object at high power. On a better night at 220X there are four stars involved in the nebula including one of about 12th magnitude on the eastern edge. Averted vision elongates the nebula; the pretty bright star involved also makes the nebula appear comet-shaped. Using direct vision, the glow of this planetary is round. An unusual effect – a "blinking plane-

This image of NGC 6804 was taken with a 20 inch R-C telescope by Adam Block/NOAO/AURA/NSF.

tary" that changes shape as you go from direct to averted vision. Try for yourself. Adding the UHC filter makes all this fun go away, but it does enhance the nebula very much.

B142-3 DRKNB AQL 19 40.7 +10 57

This fascinating area of dark nebulae is about 2000 light years distant. It is dark enough to completely block off the stars on the other side of the dark cloud. Therefore, this dark nebula got a darkness rating of 6 out of 6. Because the combination of the two nebulae forms a capital letter E, this is often called the "E nebula." It is one of the most obvious of the dark markings in the Milky Way. About 100 years ago, E.E. Barnard discovered many of these dark areas photographically. I have a fairly refined sense of humor when it comes to word play. Some of my friends call it demented, but I don't listen to them anymore. Anyway, I find it fascinating that a man with the first initials "E.E." would discover a nebula that gets the nickname "E"!

OK, enough of the trivia, the view of this area reminds me why I traded for the 4 inch (100 mm) f/6 refractor. It is an excellent RFT (Rich Field Telescope), so the short focal length provides beautiful wide field views up and down the Milky Way. The 4 inch with 22 mm Panoptic eyepiece shows that B 142 has a dark finger of material to the west that is very prominent. There are no stars involved in the western section, but there are four pretty faint stars seen in the eastern part. B 143 is an arch of darkness with the southern end more prominent. There are several thin dark fingers of material off to the north and northwest. This is a "Wow" object on a good night.

I also called this a "Wow" some years ago with the 6 inch f/6 Maksutov-Newtonian and the 22 mm Panoptic. This multibranched *dark* nebula fills the field of view with very dark lanes. No stars intrude into the darkest places. A giant "E" with a Milky Way star field on all sides. This is a fascinating area of the sky. The interplay of bright and rich star fields with the black lanes formed by this nebula draws me back to observe here season after season.

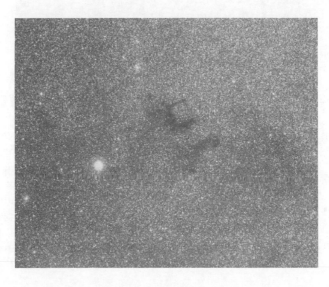

This photo of Barnard 142 and 143 was taken with an 8 inch Schmidt camera by Chris Schur.

Ara Nebulae

NGC 6188 BRTNB ARA 16 40.1 –48 40

NGC 6193 OPNCL ARA 16 41.3 –48 46

This is one of those areas of the sky that photographs better than it appears during a visual observation. It is an open cluster (NGC 6193) that has a nebula (NGC 6188) associated with it. Both of my observations are from Australia.

I first observed NGC 6188 and 6193 from Jim and Lynne Barclay's backyard near the little town of Ellesmere, Queensland. With a 5 inch (120 mm) refractor the cluster shows eight stars with a double star involved. The nebula is very faint. Averted vision helps show a line between a difficult nebulosity and a dark nebula.

At the Maidenwell Observatory with one of the three 14 inch (256 mm) SCTs and a 40 mm eyepiece the cluster is resolved into 12 stars including a bright matched double; both are pure white. The nebulosity is low surface brightness; there is a line between light and dark. Adding the UHC helps some, but this is not a bright object at all. With averted vision, I can get a hint of the dark "elephant trunks" that wind their way out into the Milky Way on all sides.

The image of NGC 6188 is by Chris Schur with an 8 inch Schmidt camera. Some of the dark areas at the bottom of the image are from trees. This object only rises 8 degrees from the latitude of central Arizona.

Cepheus Nebulae

NGC 7023 CL+NB CEP 21 00.5 +68 10

Among the information in Chapter 6, it was given that there are fewer reflection nebulae than any other type. This is because the angle must be just right to see the bright star's reflection in the dust cloud that forms the nebula. So this object was included in the observations because it is one of the brightest of the reflection nebulae. Much of that brightness comes from the seventh-magnitude star that is embedded in the nebula.

With the 13 inch at 100X, NGC 7023 is bright, large and elongated 2X1 in PA 0 degrees. This irregularly shaped nebulosity was seen immediately. A rather interesting observation: the central section does not contact the central star. There is a small dark donut around the star and then the glow of the nebula begins. The UHC filter makes the nebula smaller; this object must have a bizarre spectrum. The field has a dim glow from outer sections of the nebulosity. This nebula is in a portion of the Milky Way that is much obscured by dark nebulae.

This image of NGC 7023 was taken with a 20 inch R-C telescope by Adam Block/NOAO/AURA/NSF.

NGC 7129 CL+NB CEP 21 43.0 +66 07

The NGC number for this object is applied to both the cluster of stars and the nebula. In the 11X80 finder it is seen as a fuzzy spot. With the Nexstar 11 at 125X the cluster is pretty bright, pretty small and poor in stars. Six stars were counted in a 10 arc minute area. The nebula is pretty faint and pretty large with an irregular figure. Of the six stars in the cluster, four of them are involved within the nebula. Averted vision makes it much easier to see the nebula. Going to 200X is too much for the nebula, but I can see two very faint stars in the center that were missed at lower power. Averted vision makes the nebula much more prominent. The UHC and OIII filters do not help, but the Deep Sky filter does bring out the nebula with more contrast.

This image of NGC 7129 was taken with a 20 inch R-C telescope by Adam Block/NOAO/AURA/NSF.

IC 1396 CL+NB CEP 21 39.1 +57 30

B160 DRKNB CEP 21 38.0 +56 14

B161 DRKNB CEP 21 40.3 +57 49

B162 DRKNB CEP 21 41.1 +56 19

IC 1396 is one of the largest nebulae in the sky, so this is obviously a "busy" area. It is easy to find because the famous star Mu Cephei is on the northern edge of the nebulosity. The nebula is about 1500 light years distant and has the star cluster Trumpler 37 within it. Because it is low surface brightness, this nebula will only show off detail on an excellent night.

With the 6 inch f/6 Maksutov-Newtonian on a night I rated S = 6, T = 8, my notes show this nebula as very faint, very, very large, with an irregular figure and brightest on the north side. That observation was with the 35 mm eyepiece and a UHC filter. Removing the filter makes the nebula much smaller and more difficult to see. I counted 22 stars in the Trumpler cluster with the 14 mm Ultra Wide eyepiece. There is a triple star near the center and most of the cluster is one long chain of stars with a sprinkling of a dozen others.

In the 13 inch at 60X this nebula is very, very large, very faint and needs low power to fit it into a single field of view. There is a hint of the dark lanes with the UHC filter in place. However, I found the dark markings more obvious at 135X with the UHC installed. Barnard 161 is a comet-shaped dark marking on the north side of the cluster. It is easily the most prominent of the dark nebulae associated with this complex area. Barnard 160 and 162 are on the southern edge of this huge nebula. I cannot differentiate which is which, but there is a wide, dark lane that projects into the nebula from the south and I assume this was it.

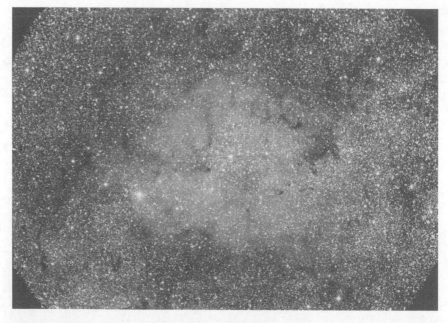

This photo of the IC 1396 region was taken by Chris Schur with his 8 inch Schmidt camera.

Corona Australis Nebulae

NGC 6726 BRTNB CRA 19 01.7 –36 53

NGC 6727 BRTNB CRA 19 01.7 –36 53

NGC 6729 BRTNB CRA 19 01.9 –36 57

Be 157 DRKNB CRA 19 02.9 –37 08

This entire field is fascinating! There are two prominent bright nebulae with stars involved. A huge dark cloud, Bernes 157, blocks off about 90 percent of the stars on the south side of the field of view, while the north side includes NGC 6723, a globular which is bright, large, much brighter in the middle and well resolved. It is as if the stars of the southern part of this area were gathered together to form the star cluster and then left behind the dark nebula.

Of course, the globular cluster has absolutely nothing to do with this amazing field of nebulae; it is many thousands of light years in the distance. As far as my note keeping is concerned, NGC 6723 is actually in the constellation of Sagittarius. The nebulae are about 500 light years away and are obviously in front of the dark nebula. If I had control of the starship *Enterprise* for a month, this would be one of the places I would travel. Let's see what I have for notes on the individual parts of this rich and complex field of view.

NGC 6726 in the 13 inch is easy at 100X without a filter; I see it as a seventh-magnitude star with a pretty faint, pretty large haze surrounding it. NGC 6726 and NGC 6727 are seen as one object with a "figure 8" shape at low power. At 220X and a moment of good seeing, there is a dark lane that will appear to separate the two nebulae.

NGC 6729 contains two stars, one of eighth magnitude and the other of ninth magnitude. They are within a pretty faint, pretty large nebula. The nebula is elongated 1.5X1 in PA 45. The UHC filter does not help the contrast of either of these nebulae.

These previous observations were taken on one of the best nights I have ever had. A.J. Crayon and I have been going out observing for many years in Arizona. We have found that a rain shower on Tuesday or Wednesday will clear out the air and provide excellent transparency for the weekend when we wish to go out observing. This was one of those weekends, at high altitude with the clearest of skies. I rated the seeing at 7/10 and the transparency at 10/10 (perfect). A.J. and I both said that if we even get to the top of Mauna Kea we will call it an "11" – shades of *Spinal Tap*!

From Australia, NGC 6729 passed almost directly overhead. In the 12 inch it was a pretty bright, pretty small, comet-shaped nebula with a 12th-magnitude star in the tip at 135X. There is a faint star on the south side. The UHC filter dims this object quite a bit; that is to be expected when viewing a reflection nebula

The Bernes 157 is one of the most prominent dark nebulae I have ever seen. In the 12 inch with a 30 mm eyepiece from Australia I only saw 15 stars in a field of view of about 25 arc minutes wide. The fact that 12 inches of aperture only shows 15 stars in a field of view that is in a constellation right next to Sagittarius must mean that there is a really dark nebula blocking off the stars on the other side.

This image of the busy region of Corona Australis was taken by Jon Christensen with a Takahashi refractor. The globular cluster in NGC 6723, the brightest part of the nebula with two pretty bright stars involved is NGC 6726-7. The comet-shaped glow that is below and left is NGC 6729 and the dark nebula is Bernes 157.

Cygnus Nebulae

NGC 6826 PLNNB CYG 19 44.8 +50 32

NGC 6826 is the Blinking Planetary. If you look directly at the planetary the central star is prominent compared to the greenish nebulosity. Then averted vision will make the nebula appear brighter and overwhelm the star. Alternating between direct and averted vision will produce a blinking on-then-off effect that is fascinating. There are very few telescopic views that "do something" when you view them, so don't miss this interesting nebula. There are several other planetaries that have the proper ratio of star brightness to nebula brightness to show the effect, but none are as obvious as this one.

In the 6 inch f/6 Maksutov-Newtonian with an 8.8 mm eyepiece on a good night ($S = 6$ and $T = 7$), this planetary was pretty bright, pretty large and round. It showed a stellar nucleus and was very light green in color. Raising the power with a 5 mm Lanthanum eyepiece will show off the blinking effect quite well. With averted vision it grows 1.5 times its size with direct vision.

Using Dave Fredericksen's 12.5 inch f/6 Newtonian on an excellent night, this planetary is bright, large, somewhat elongated and greenish. The central star is easy at 180X as it floats in a rich Milky Way field of view. The blinking effect is easy at this power. Look straight at it and you see just a tiny nebula around the central star – look away and it blows up to at least three times its size with direct vision, a mesmerizing effect.

In Jim Christensen's 25 inch scope at the Texas Star Party, this planetary is a sparkling neon green in color and the light blue central star is seen 100 percent of the time, so this is *not* the "blinking" planetary with a large aperture scope. Using an 11 mm eyepiece, it is bright, pretty large, round and the central star is unmistakable. There are very subtle brighter areas within the nebula; they are called FLIERS – fast low ionization emission regions. Going to a 5 mm eyepiece (635X) shows two low surface brightness oval spots at 10 and 4 o'clock from the central star. This is the bipolar material being ejected from the middle of this nebula.

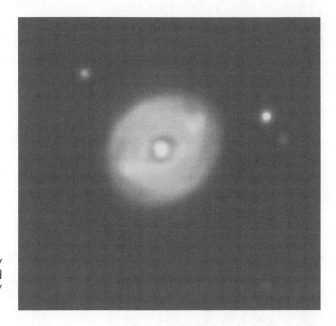

Image with 20 inch R-C by
Tom Boerner and David
Young/Adam Block/NOAO/
AURA/NSF.

NGC 6960 SNREM CYG 20 45.6 +30 43

This is visually the most prominent supernova remnant in the sky. It is the gas and
dust that an exploding star spewed out into space about 30,000 year ago. In binoculars the Veil Nebula is easy to see on a good night.

    This is a fascinating region in most any telescope you can muster. With a small
telescope, the wide field view will show you the beautiful loops of glowing nebulosity that fill this portion of the sky. This is a rich, starry region and the addition
of this bright supernova remnant adds to the beauty of the view. If you have a larger
scope then you are not going to be able to fit any entire arc of the Veil into your
telescope, regardless of the advertisements from eyepiece makers. But you can see
plenty of detail within the nebula as the swirling gas tendrils form a cosmic "taffy
pull" effect.

    NGC 6960 is the portion of the Veil Nebula that is passing behind the bright star
52 Cygni. In the 4 inch it is not as easy as the other section of the Veil (NGC 6992).
The star makes it somewhat more difficult to see the streamer of nebulosity at low
power. Using the 22 mm Panoptic and a UHC filter makes it easier to see; the star
is dimmer through the filter and the nebula is enhanced. The northern "pointy end"
of this part of the Veil is higher in surface brightness than the southern "split" end.

NGC 6992 SNREM CYG 20 56.3 +31 42

Using the 22 mm Panoptic and a UHC filter provides a great view of this famous
object. This entire section of the Veil fits into the field of the 4 inch RFT with room
to spare. It appears as a very elongated glowing streamer within a rich field of stars.
There are 12 stars involved within the nebula. This side of the Veil Nebula appears
to me like the curved blade of a sickle.

    In the Nexstar 11 with a 27 mm eyepiece and a UHC filter the curved strands
that are so prominent in photographs of this object are easy to see. As the scope is

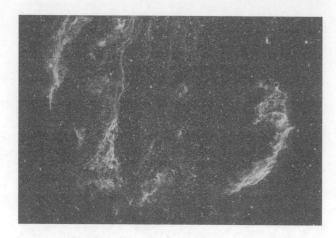

This image of the Veil Nebula is by Jon Christensen with a Takahashi refractor.

scanned along the length of NGC 6992 there are bright regions of the nebula that are areas of this supernova remnant that are more ionized than fainter areas. Both the UHC and OIII filters do an excellent job of enhancing the nebula.

NGC 7000 BRTNB CYG 20 58.8 +44 20

This is one of the most easily recognized deep sky objects on photographs. Many a beginning astrophotographer shot a wide field photo of Cygnus and when the film was developed was pleased to find that they "got the North America Nebula" on the image. Me too.

It is seen as a naked-eye glow near Deneb on a good dark night. I really doubt that we are seeing the nebulosity; there are lots of stars in this area along with the nebula and I think it all blends together. Regardless, this is a great place to use a pair of binoculars. Once you are well dark-adapted, the form of this nebula stands out nicely even with my small 8X42 binoculars. The glow is almost 2 degrees in size and a quite prominent dark lane, Lynds Dark Nebula 935, forms the East Coast. The "Mexico" region is also easy to spot; it has good contrast with higher surface brightness compared to the rest of the nebula.

This is an object that reminds me of the fun you can have when sharing the sky with friends. On a trip to dark skies some years ago, Rick Rotramel showed up at a SAC star party with his 11X80 binoculars and a new parallelogram mount. It made those binoculars much easier to use. The view of the North America Nebula was excellent; the shape was just framed in the field of the 11X80s. There was good contrast between the dark lanes and the bright glow of the nebula. There were lots of stars involved within the glow and the star density to the north of the nebula was fascinating – an ultra-dense field of stars. If you get a chance to use a large pair of binoculars, take it.

In the 4 inch refractor and a 35mm eyepiece with the 2 inch UHC filter this is a great view of that famous shape. It is a bright, very, very large, very irregular figure (like North America!) and shows 41 stars involved. The dark lane on the east side that cuts off the nebula is very prominent and has an orange star involved. The star cluster NGC 6997 is about at the "Detroit" area and it shows a dozen pretty bright stars, but it is somewhat scattered. The Pelican Nebula (IC 5070) is easily seen at the edge of the field.

This image of the North America Nebula is by Chris Schur with an 8 inch Schmidt camera.

NGC 7048 PLNNB CYG 21 14.3 +46 17

NGC 7048 is a pretty faint, pretty small and elongated planetary that looks like a grey blob in the Milky Way at 135X in the 17.5 inch. Using the 13 inch on a good night I said it was pretty bright, pretty large and somewhat brighter in the middle at 220X. This object could be recognized at 100X, but higher power brought out some detail. It is irregularly round and very light green. The Milky Way field is very rich in this area. It would seem that the quality of the evening has a lot to do with the perceived brightness of this object.

This image of NGC 7048 is from Chris Schur with a 12.5 inch f/5 Newtonian.

B168 DRKNB CYG 21 53.2 +47 12

IC 5146 CL+NB CYG 21 53.4 +47 16

The Cocoon Nebula and its associated dark lane have been photographed many times. I would say that it is one of those objects that any visual view I have ever had does not show me the detail seen in a good astrophotograph of this region. But that does not mean you should skip over it. Every observing list needs a few challenge objects.

With the 4 inch f/6 RFT and a 22 mm Panoptic, the thin dark lane B 168 is easy. The Cocoon Nebula is extremely faint with this small aperture and the UHC filter does not help. I only suspected a very faint glow at the end of the dark lane with averted vision. That is with a 22 mm Panoptic eyepiece. I tried the 14 mm UWA and the Cocoon Nebula was somewhat easier, but not much. That is a difficult nebula in a small aperture, even on a great night. All this detail is in a very rich Milky Way field of view.

Using the Nexstar 11 and a 27 mm Panoptic on a great night, I saw the Cocoon as faint, large, irregularly round and including five stars. Averted vision helps the contrast a lot. There is a hint of a thin dark marking within the nebula. The Cocoon Nebula is at the end of a pretty easy dark lane. I rated this excellent evening at S = 7, T = 9. The UHC filter did not help the contrast of this object. I have read an observation from noted deep sky observer Steve Gottlieb who said that he saw an increase in contrast with the H Beta filter. That may be what I will buy with the royalty payments from this book.

This wide-field shot of the Cocoon Nebula and the dark lane that leads to it (Barnard 168) were shot with a Takahashi refractor by Adam Block/NOAO/ AURA/NSF.

This closeup image of the Cocoon Nebula (IC 5146) was taken by Chris Schur with a 12.5 inch f/5 Newtonian.

Lyra Nebulae

M 57 PLNNB LYR 18 53.6 +33 02

The Ring Nebula is probably the most famous planetary nebula in the sky. Using the 4 inch and a 27 mm Panoptic eyepiece it is easily seen as a nonstellar dot. Even though it is tiny, there is an elongated dot in a pretty rich field of view. Raising power with the 8.8 mm makes this an obvious ring. The annular disk is elongated 1.8X1 with a dark center that is more obvious with averted vision.

The Nexstar 11 at high power will reveal a thin fog of material that fills the Ring. John Herschel said this appeared to him like "gauze over a hoop" and I agree. With a 6.7 mm eyepiece the effect is obvious. The central star is never seen as a steady dot of light in the middle of the Ring, but it does wink at me occasionally. I would estimate that I see the central star 5 percent of the time. Adding the OIII filter really enhances the annular glow of the Ring, but the central star is no longer seen with the filter in place.

Over 20 years ago the Saguaro Astronomy Club helped the staff at Kitt Peak Observatory pass legislation in Arizona that limited outdoor lighting. As a reward for our help we were provided two hours of observing time on a 36 inch f/7.5 telescope. A more eager group of observers never existed, including the student telescope operator; he had never had a chance to just observe with the telescope. We realized that the field of view would be small, so we chose objects to observe that would fit, and the Ring was first on the list. The old 25 mm Kellner eyepiece that was used every night to get the photometer aligned on target was a mess, but even then we knew it was going to be a great view. We replaced the old eyepiece with my brand-new 16 mm research grade Erfle. That put the scope at 427X and the Ring Nebula was at its best. The annular structure was a medium green color and the central star was held steady about 40 percent of the time. The nebulosity within the ring shape was easy and really became enhanced with averted vision. In moments of good seeing there were thin fingers of nebulosity that pointed inward toward the central star. The ring shape is about one third of the field of view in this telescope. We called it the "Giant Green Donut." It was a memorable night if ever there was one.

This image of the Ring Nebula is by Chris Schur with a 12.5 inch f/5 Newtonian.

Ophiuchus Nebulae

B 59, 65-7 DRKNB OPH 17 21.0 –27 00

If you have not spent much observing time on dark nebulae, here is a great place to start. If you travel to, or live under (lucky you), fairly dark skies then this is a naked-eye dark marking just north of the Large Sagittarius Star Cloud. It does indeed look like a dark smoking pipe. From northern temperate regions it appears to show the bowl to the left and the stem to the right, toward Scorpius. There are several Barnard numbers associated with this area, but the Pipe itself is mainly B 59. The LDN stands for the Lynds Dark Nebula catalog and the entire Pipe Nebula is number 1773 in that catalog from Dr. Beverly Lynds.

Because this is a large dark nebula it demands a wide field of view to enjoy this area at its best. After some hesitation I decided to purchase a very good pair of binoculars. I had been living with the 75 dollar pair of 10X50s for many years. Then I dropped them and it seemed like a good time to reconsider my position. It took a lot to get me to spend 300 dollars on a pair of 8X42 binoculars, but I bought the Orion Savannah 8X42 binoculars and I have been happy ever since. I am also motivated to be more careful with these.

Using those 8X42 binoculars, this field of view is absolutely fascinating. The interplay of light and dark across this part of the sky really does make it look like the dark nebulae are in front of the star fields beyond. There is a curved line of stars that outlines the stem of the Pipe and then the dark inner section is contained within these eighth- and ninth-magnitude stars. There are several places where dark fingers stretch out from the stem of the Pipe and meander into the rich star fields nearby.

The 4 inch f/6 RFT was made for views exactly like this. With a 22 mm Panoptic this is a great view, and there is lots of contrast between the dark markings and the starry background. The stars that outline the stem of the Pipe are easy and the darkness is composed of several dark areas near each other. There are dark lanes that extend north and more prominently to the south of the stem feature. B 59 is the tip of the stem of the Pipe. I see it as dark, large, somewhat elongated N-S: that is, "across" the stem of the Pipe.

This montage image of the Pipe Nebula is by Chris Schur with an 8 inch Schmidt camera.

B 63 DRKNB OPH 17 16.0 –21 23

Some years ago, when he was the editor of *Astronomy* magazine, Richard Berry looked at a photograph of this region and said "Looks like a dark horse." Sure

This is a photo of the entire "Dark Horse"; the Pipe Nebula is on the left and it is the rear leg. The prancing leg is Barnard 63, on the right of this frame. This photo was taken by me; it is a 15 minute exposure with a 135 mm lens.

enough, with a pair of binoculars from a dark site you can see the Pipe Nebula as a back leg, the Snake (B 72) as a small navel or umbilical cord and this area has two front horse legs; this one is lifted as if the horse is prancing.

With the 4 inch f/6; on a night I rated S = 6, T = 7, this is a large oval elongated darkness. Using my new 13 mm Orion Stratus eyepiece shows in the central region there are *no* stars in an area about one degree by half a degree, an astonishing view. The entire area of the Dark Horse is a fascinating and contrasty view of light and dark.

Sagittarius Nebulosity

M 8 CL+NBSGR 18 03.7 –24 23

The Lagoon Nebula is an easy target in a fascinating Milky Way field. Virtually any optical aid will provide plenty of detail as you look toward the center of our Galaxy. M 8 is about 5000 light years distant.

The Lagoon Nebula is an obvious naked-eye bright spot in the Milky Way, even on a mediocre evening. In 8X25 binoculars this area in spectacular. The Lagoon and Trifid nebulae fit in the same field with star chains and dark lanes winding their way through the entire field of view. The little binoculars show three stars resolved within the glow of the Lagoon and it is elongated 2X1 E–W. The eastern end of the Lagoon is brighter. Using 10X50 binoculars shows eight stars resolved, half of them clustered on the eastern side. Averted vision makes the nebula grow much larger. The dark Rift in the Milky Way is obviously in the background of this glowing cloud.

On a night far from the city lights, using the 13 inch at 60X, M8 is a very bright, very, very large and somewhat compressed cluster with lots of nebulosity. I said that I would use some of the proceeds from my first book for a 35 mm Panoptic eyepiece and this is the first night with that new toy. A 1 degree view shows the nebula is 80 percent of the field of view and the dark lane that gives the Lagoon its name is obvious. There are 40 stars included within the cluster and there are

This image of the Lagoon Nebula (M 8) is by Chris Schur with a 12.5 inch f/5 Newtonian.

another 50 stars among the outer parts of the nebula; 10 of the stars are within the dark lane. There are two prominent ninth-magnitude stars across the dark Lagoon feature from the cluster; the southernmost of the two is 9 Sgr. Moving the scope just west of that star is the brightest section of the nebulosity, it is called the Hourglass. That nickname is appropriate, as at 200X the glow is indeed in the shape of an hourglass.

Adding the UHC filter makes the nebula grow 1.5 times in size. However, I don't like the view with the UHC filter because it dims the stars involved within that lovely glow and that is a large part of the beauty of this area of the sky. Just my opinion.

M 20 CL+NB SGR 18 02.7 –22 58

My friend Curt Taylor passed on to the big observatory in the sky several years ago; he was an avid observer of the Moon and planets. Occasionally, I would coerce Curt out of town to view some deep sky stuff with me. When the 6 inch f/6 Maksutov-Newtonian was brand new we traveled 50 miles from Phoenix to see what it could do, and the Trifid was one of the nebulae that we viewed. In the 6 inch with a 14mm UWA eyepiece and no filter, the Trifid shape is obvious. There are 10 stars

This image of the Trifid Nebula (M 20) is by Chris Schur with a 12.5 inch f/5 Newtonian.

This image of the region of M8 and M20 is by Chris Schur with an 8 inch Schmidt camera. The Lagoon Nebula (M 8) is the large, extended nebula at the bottom of the photo. The Trifid Nebula (M 20) is near the middle of this image. Note the profusion of dark lanes in this area of the Milky Way.

involved in the nebula and the dark lanes are easy within this bright, large, irregular figure. The double star HN 40 (HN = Herschel Number) is an easy split. At 165X the star becomes a triple and shows two yellow stars and one light blue star embedded within the Trifid nebulosity.

Remember that there are two sections to the Trifid Nebula: a southern emission region that has the dark lanes in it and the north side reflection nebula. Adding the UHC filter cuts the size of the reflection nebula in half and makes it much fainter. However, the emission nebula has much more contrast with the filter. The pretty bright star within the reflection nebula is light orange. The emission and reflection sections have a different "texture." The reflection region is smooth and the emission portion is rough, like the lumps, or hummocks, in the lunar terrain. Reminds me of Curt when we were viewing the Moon.

M 17 CL+NB SGR 18 20.8 –16 11

This nebula has always been a favorite of mine. When Sagittarius, Scorpius, Cygnus and Aquila are above the horizon there are lots of targets to view, but I find myself returning here often. There is much to see in and around M 17. As far as nebulae go, only the Orion Nebula (M 42) and the Eta Carinae Nebulae (NGC 3372) are brighter than M 17. Also, as you change telescopes, eyepieces, and filters the view of this fascinating object changes its outline to reward you with a variety of differing silhouettes. The NGC description includes "eiF," which means "extremely irregular Figure." This is the reason that several different nicknames have been applied to this object. So here is your opportunity to take lots of notes. Spend some time with this object and really take in its beauty.

On a good night that I rate 6/10 for both seeing and transparency, I am using the Celestron Nexstar 11 to test the effect of several types of filters on M 17. With the 22 mm Panoptic and no filter the bright "checkmark" section is obvious, and there are two dark lanes that cut across the bar feature. There are two patches of faint nebulosity that are part of the outer loop of nebulosity, but I don't see the entire "Omega" feature. Adding the wideband Deep Sky filter brings out much of the nebulosity. This makes the nebula more contrasty, and the dark lanes through the checkmark or swan feature are much more prominent. The fainter arch of nebulosity that creates the Omega or horseshoe shape stands out nicely against a

This image of M 17 was taken by Chris Schur with a 12.5 inch f/5 Newtonian.

much darker background, all without changing the star colors in the field. I like that. The UHC filter shows a similar amount of nebulosity to the Deep Sky, but the contrast of the background is better. The UHC does get rid of many of the field stars. It is certainly better contrast, but you give up a lot. Now on to the OIII filter and the field is now *black*, there are a lot less stars and less nebulosity. The nebula sections that are seen are very contrasty, but the faint outer sections of the nebula are not nearly as large. Remember, we are having fun here; there is no "correct" view.

With Tom Clark's Yard Scope, a 36 inch f/5 Dobsonian on an excellent night that I rated S = 7, T = 8, this beautiful nebula is stunning. With a 20 mm eyepiece and UHC filter it shows at least as much detail as any photo from the 200 inch on Mt. Palomar. There are 27 stars involved and dark lanes cut the bright "checkmark" into sections; many bright pieces of the nebula extend far beyond the center of the nebulosity out into the Milky Way. Two lovely delicate chains of stars are located in the "head of the Swan"; they include a beautiful orange star that I called "the Eye of the Swan."

During my trip to Australia it took me a few minutes to find M 17! I am so used to the lineup of objects going upward from the horizon – Large Sagittarius Star Cloud, Lagoon Nebula, Small Star Cloud, and then M 17. But, from Queensland, this nebula is *below* the Small Sagittarius Star Cloud and it took me two or three minutes to "discover" that orientation.

After I put M 17 into the 12 inch f/15 Cassegrain with a 30 mm eyepiece, the view was excellent. The dark markings on the body of the Swan looked like a barber pole. Inserting the UHC filter shows the faint outer reaches of the nebulosity are now larger than the field of view on all sides. A dedicated observer can spend lots of time here following all the loops and filaments of nebulosity that trail off beyond the field of a low-power eyepiece.

NGC 6445 PLNNB SGR 17 49.3 –20 01

Using the 6 inch Newtonian and a 22 mm Panoptic eyepiece, this nebula is pretty faint, pretty small, round and not brighter in the middle. This planetary is "above" a globular cluster of about the same brightness, but the globular is twice its size. Moving to the 8.8 mm eyepiece still shows it as pretty faint, but now it has some size and is elongated 2X1 in a PA of 135 degrees.

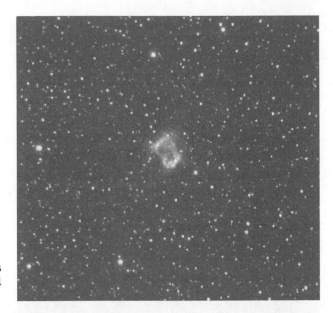

This image of NGC 6445 is
by Jim Barclay with a 14
inch SCT.

This planetary is pretty bright, pretty large and has an elongated box shape at
200X in the 13 inch. The outer rim of this planetary is brighter than the center. A
white and blue double star is nearby. When the power is raised to 330X, the scope
shows two "polar caps" on either side of an elongated shape (1.5X1 in PA 135).
Pierre Schwaar called this planetary the "Micro Dumbbell Nebula." It does have
some resemblance to M 76, the Little Dumbbell, in Perseus.

B 84 DRKNB SGR 17 46.5 –20 11

Barnard 84 and its surrounding area is a rich star field if ever there was one! In
the 4 inch RFT refractor and a 22 mm Panoptic the view is awash with star points.
B 84 itself is an obvious dark marking in front of all those stars. I see it as irregularly round with a pretty bright star at the northern edge. The compact globular
cluster NGC 6440 is at the edge of the field of view.

In the Nexstar 11 with a 27 mm Panoptic this dark area is an oval shape with a
matched double star just off the center of the oval. The star at the edge of the dark

This image of Barnard 84 is
by Jon Christensen with a
Takahashi refractor.

area is light yellow and across from that star, there are several thin dark lanes that make their way north and west out into the rich Milky Way star glow.

B 86 DRKNB SGR 18 02.7 –27 50

B 86 is a *dark* Barnard Nebula right next to the open cluster NGC 6520. This dark marking has an hourglass shape and is about 5 minutes in size using the 13 inch at 100X. There is a lovely orange star in the field. Don't miss this very nice area. Because it is such an obvious dark oval in the Milky Way, I have heard it called the Ink Spot for many years.

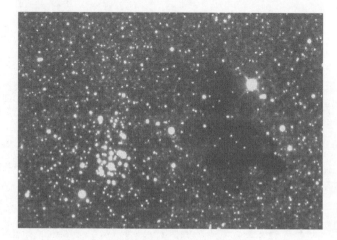

This image of B 86 is by Chris Schur with a 12.5 inch f/5 Newtonian.

Scorpius Nebulae

NGC 6337 PLNNB SCO 17 22.3 –38 29

This is a low surface brightness planetary in a rich field near the Stinger stars of Scorpius.

With the 4 inch f/6 and a 14 mm eyepiece with the UHC filter, the nebula comes and goes with the seeing at this aperture. It can be held about 50 percent of the time, a tiny low surface brightness dot in the Scorpius Milky Way. This is on a night I rated the seeing at 6/10 and the transparency the same.

On a somewhat better night with the Nexstar 11, this nebula is pretty small and faint with an irregular figure. A faint fuzzy even at low power, averted vision brings it out. That observation was with no filter and the 22 mm eyepiece. Moving up to the 14 mm with no filter shows this object as two lobes. The left one has a star involved; the right side is fainter, with a brighter spot involved that is much more obvious than the rest of the nebula. The connection between the two lobes is just suspected without the filter. Adding the UHC makes all the difference; now the nebula is annular and that ring shape is seen and held steady with direct vision. The right side is brighter, but the nebula is pretty faint overall. The star involved in the left side is now gone since the filter raised the contrast of the nebula.

This image of NGC 6337 is by Jim Barclay with a 14 inch sCT.

The 32 inch f/4 at Jay LeBlanc's observatory shows this off as a lovely annular planetary. The big aperture shows two pretty bright stars in the annulus and one faint central star involved within the dark center of this little ring.

IC 4628 BRTNB SCO 16 57.0 –40 27

This nebula is involved with one of the most entrancing areas of the sky in my opinion. I have heard it called the "Table of Scorpius" for a decade, from the naked-eye appearance of a large central table leg, like old wooden table designs. It includes several open clusters of a wide variety of types, a dark nebula and this faint emission nebula at the northern edge. There is plenty to see in this rich area of the sky.

In the 4 inch f/6 RFT on a great night (S = 7, T = 8) with a 22 mm Panoptic eyepiece, this nebula is just barely seen without a UHC filter; it is averted-vision only. Adding the UHC makes the nebula easy, but still pretty faint, very large, elongated 2.5X1 with 16 stars involved. Going to the OIII filter adds to the contrast of the nebula, but really darkens the sky and gets rid of many of the stars. Barnard 48 is a dark nebula just to the south of IC 4628; it cannot be seen in the 4 inch without a filter because there is no contrast. Adding the UHC makes it stand out much better. Now the south side of the glowing nebula stands out better and it is cut off by the elongated dark lane on that side.

Viewing the Table of Scorpius in the 11X80 binoculars from Australia is amazing. The big binoculars encompass this area perfectly, from Zeta 1+2 to Mu 1+2 Sco. NGC 6231 is a very compressed open cluster: 12 stars are resolved in the cluster, another 20 stars form a "spray" of stars around it. Trumpler 24 is a more scattered cluster to the north; it shows 12 stars with a dark lane down the middle. Barnard 48 is seen as a very star-poor area. The nebulosity is not seen with the big binoculars when I am holding them by hand only. I lean over the opening in the dome and brace my elbows on the lip and now the nebula is just barely seen as a faint haze to the north of Trumpler 24. Aussies call this area the "False Comet," and I agree, as with the naked eye it does look like a fourth-magnitude comet with a bright head.

This image of NGC 4628 and the rest of the Table of Scorpius was taken using a Televue 76 refractor, by Adam Block/NOAO/AURA/NSF.

Vulpecula Nebulae

M 27 PLNNB VUL 19 59.6 +22 43

The Dumbbell Nebula is the brightest planetary in the Messier list and one of the brightest nebulae in all the sky.

With the 4 inch and a 27 mm Panoptic it is easily seen as a small, pretty bright box-shaped nebula. This object has much higher surface brightness than the Ring Nebula. Using an 8.8 mm eyepiece the "apple core" or Dumbbell shape is immediately seen; the small aperture only shows two stars involved. Averted vision shows some of the faint outer nebulosity beyond the bright Dumbbell shape.

Moving up in aperture to the Nexstar 11 and a 14 mm eyepiece shows seven stars involved, with the last two extremely faint. The outer layers of nebulosity are pretty easy on a good night; averted vision makes them unmistakable. The nebula is bright, pretty large, elongated 1.5X1 with a high surface brightness core and fainter glow around it. This nebula has a faint green color. Adding the UHC or the OIII filter shows the outer nebula with much greater contrast. The filter actually makes the Dumbbell Nebula almost round by filling in the central sections.

Many years ago at the Riverside (California) Telescope Makers' Conference, the folks from Questar drove across the US and set up a Questar 12, a beautifully made 12 inch f/15 Maksutov. Like many a big Maksutov, it took quite a while to cool down.

However, by midnight the views were excellent. At 180X with a 15 mm eyepiece the Dumbbell Nebula was a beautiful light green color that really convinced you that the thing you were seeing was a glowing gas cloud. There were 12 stars involved within the nebula and showed the nebulosity to have a mottled texture, somewhat like the Orion Nebula around the Trapezium. These light and dark markings were seen across the face of the entire nebula. A memorable view with a memorable telescope.

This image of M 27 is by Chris Schur with a 12.5 inch f/5 Newtonian.

Chapter 12

Nebulae are Just the Start

Well, here we at the end of my book on nebulae. This certainly was fun writing and I hope you had fun reading it.

I do have one important point to make and that is the fact that the abilities you have heard about in this book will be useful for all of your observing. Don't think that averted vision, for instance, is only useful when observing nebulae. One great example of this is globular clusters: as you move your eye from direct to averted vision the view of many globulars really changes. Direct vision will give you the best view of the detail within the core and averted vision will bring out those lovely looping chains of stars that make these giant star clusters huge in the telescope.

That is just one example; I will leave you to explore others. But do remember to make an observing list, use a variety of magnifications and take some notes about what you have seen. If you do those things then I promise you will get more from your observing sessions.

And this is supposed to be fun. Being an amateur astronomer is not a contest or a race, just an opportunity for you to see the Universe in all its glory.

Clear Skies to us all;

Steve Coe

Appendix

Object	Other	Type G+C+N	Con	R.A.	Dec.	Mag	SIZE_MAX	SIZE_MIN	Class	BRSTR	Description and notes
NGC 206		G+C+N	AND	00 40.5	+40 44	none			Pec		vF, vL, mE 0 degrees; neby in south end of M 31
NGC 7662	PK 106-17.1	PLNNB	AND	23 25.9	+42 32	8.6	17 s	14 s	4(3)	14	!!! Planetary or annular neb, vB, pS, R; Blue Snowball Nebula
PK 308-12.1	He2-105	PLNNB	APS	14 15.5	−74 13	12	35 s				
PK 315-13.1	He2-131	PLNNB	APS	15 37.2	−71 55	11.8	4.9 s			10.9	
B127, 129-30		DRKNB	AQL	19 01.6	−05 26	none	20 m	5 m	5 lr		Curved; lying north of 12 Aql
B132, 328		DRKNB	AQL	19 04.1	−04 28	none	16 m	8 m	6 lr		40 d north preceding Lamba AQL
B133	LDN 567	DRKNB	AQL	19 06.1	−06 50	none	10 m	3 m	6 CoG		On Scutum star cloud 2 deg south of Lamba Aql
B134	LDN 531	DRKNB	AQL	19 06.9	−06 14	none	6 m		6 CG		1.4 deg south of Lamba Aql
B135-6	LDN 543	DRKNB	AQL	19 07.4	−03 55	none	50 m	30 m	6 lr		1 deg north following Lamba Aql
B137-8	LDN 581	DRKNB	AQL	19 15.6	+00 13	none	180 m		3 lr		Long; curved lane on Scutum star cloud
B139	LDN 627	DRKNB	AQL	19 18.1	−01 28	none	10 m	2 m	5 EG		At south tip of B137-8
B142-3	LDN 619	DRKNB	AQL	19 40.7	+10 57	none	80 m	50 m	6 lr		Narrow lanes 3 deg north preceding Altair
B334, 336-7		DRKNB	AQL	19 36.8	+12 27	none	40 m	5 m	4 lr		2 deg north preceding B142-3
B335	LDN 663	DRKNB	AQL	19 36.9	+07 34	none	4 m		6 EG		3.5 deg preceding and 1.2 deg south of Altair
IC 4846	PK 27-9.1	PLNNB	AQL	19 16.5	−09 03	12	2 s		2	13.7	stellar
NGC 6741	PK 33-2.1	PLNNB	AQL	19 02.6	−00 27	12	9 s	7 s	4	14.7	Planetary, stellar; Phantom Streak Nebula
NGC 6751	PK 29-5.1	PLNNB	AQL	19 05.9	−06 00	12	20 s		3	13	pB, S; Annular

				RA	Dec	mag	size	size	class	mag	description
NGC 6781	PK 41-2.1	AQL	PLNNB	19 18.5	+06 32	11.8	111 s	109 s	3b(3)	16.9	F, L, R, vsbM disc; opposite dark lane from Altair
NGC 6790	PK 37-6.1	AQL	PLNNB	19 23.0	+01 31	11.4	2 s		2	16.1	B, eS, stell = 9.5 m
NGC 6803	PK 46-4.1	AQL	PLNNB	19 31.3	+10 03	11	4 s		2a	14.	stellar
NGC 6804	PK 45-4.1	AQL	PLNNB	19 31.6	+09 14	12.4	63 s	50 s	4(2)	14.1	cB, S, iR, rrr
PK 31-10.1	M3-34	AQL	PLNNB	19 27.1	-06 35	12.4	6.0 s	5.1 s	2	14.6	
PK 32-2.1	M1-66	AQL	PLNNB	18 58.4	-01 04	13	<5 ? s		1		
PK 37-3.2	Abell 56	AQL	PLNNB	19 13.1	+02 53	12.4	188 s	174 s	4		
PK 39-2.1	M2-47	AQL	PLNNB	19 13.6	+04 38	13	9.7 s	6.9 s	2		
PK 45-2.1	YY 2-2	AQL	PLNNB	19 24.4	+09 54	12.7			1	13.7	
PK 47-4.1	Abell 62	AQL	PLNNB	19 33.3	+10 37	13	161 s	151 s	2c	18.2	eF, pS, E, nBM at 165X, 2 vF* invol
PK 52-2.2	Merrill 1-1	AQL	PLNNB	19 39.1	+15 56	11.8	3 s		4		pF, vS, R, BM at 165X, averted vision helps
PK 52-4.1	M1-74	AQL	PLNNB	19 42.3	+15 09	12.9	9 s		1	12.9	
NGC 7009	PK 37-34.1	AQR	PLNNB	21 04.2	-11 22	8.3	28 s	23 s	4(6)		!!! vB, S, elliptic; Saturn Nebula
NGC 7293	PK 36-57.1	AQR	PLNNB	22 29.6	-20 50	6.3	960 s	720 s	4(3)	13.5	!, pF, vL, E or biN; Helical Nebula

!	remarkable object	!!	very remarkable object	E	elongated	s	suddenly
am	among	n	north	e	extremely	s	south
att	attached	N	nucleus	er	easily resolved	sc	scattered
bet	between	neb	nebula, nebulosity	F	faint	susp	suspected
B	bright	P w	paired with	f	following	st	star or stellar
b	brighter	p	pretty (before F, B, L, S)	g	gradually	v	very
C	compressed	p	preceding	iF	irregular figure	var	variable
c	considerably	P	poor	inv	involved	nf	north following
Cl	cluster	R	round	irr	irregular	np	north preceding
D	double	Ri	rich	L	large	sf	south following
def	defined	r	not well resolved	l	little	sp	south preceding
deg	degrees	rr	partially resolved	mag	magnitude	11m	11th magnitude
diam	diameter	rrr	well resolved	M	middle	8 . . . 13	8th mag and fainter
dif	diffuse	S	small	m	much	9 . . . 13	9th to 13th magnitude

Object	Other	Type	Con	R.A.	Dec.	Mag	SIZE_MAX	SIZE_MIN	Class	BRSTR	Description and notes
IC 1266	PK 345-8.1	PLNNB	ARA	17 45.6	-46 05	12.3	13 s		4	11.1	stellar, gaseous spectrum-Pickering
IC 4642	PK 334-9.1	PLNNB	ARA	17 11.8	-55 24	12.4	15 s		4	13.6	stellar
NGC 6188	ESO 226-EN19	BRTNB	ARA	16 40.1	-48 40	none	20 m	12 m	E+R		F, vL, vlE, B* inv; Cluster NGC 6193 and triple* h 4876 involved
NGC 6326	PK 338-8.1	PLNNB	ARA	17 20.8	-51 45	12	15 s	10 s	3b	13.5	pB, vS, R
PK 336-6.1	Peimbert 14	PLNNB	ARA	17 06.3	-52 27	12.6	8 s	6 s		14.8	
PK 342-6.1	Canon 1-4	PLNNB	ARA	17 27.9	-46 56	12.9	<10s				
PK 342-14.1	Shapley 3	PLNNB	ARA	18 07.4	-51 03	11.9	36 s				
B 26-8		DRKNB	AUR	04 55.2	+30 35	none	20 m		6Ir		Several small clouds adjacent to AB Aur
B 29		DRKNB	AUR	05 06.2	+31 44	none	10 m		6C		1.2 deg south and 2 deg following Z Aur
B 34		DRKNB	AUR	05 43.5	+32 39	none	20 m		4CG		2 deg preceding cluster M 37
IC 405	LBN 795	BRTNB	AUR	05 16.5	+34 21	10	50 m	30 m	E		* 6.7w pB, vL, vL neb: Flaming Star Nebula; variable AE Aur in center of neby
IC 410	NGC 1893	BRTNB	AUR	05 22.7	+33 25	7.5	11 m		E		Dif, many st inv; Incl cluster NGC 1893
IC 2149	PK 166+10.1	PLNNB	AUR	05 56.4	+46 06	10	12 s	6 s	3b(2)	11.3	S, vB
NGC 1931	OCL 441	CL+NB	AUR	05 31.4	+34 15	10.1	3 m	3 m	13 p n:b	11.5	vB, L, R, B*** in M; contains triple ADS 4112
NGC 1985	PK 176+0.1	BRTNB	AUR	05 37.8	+31 59	12.5	0.7 m		R	13.5	cF, S, R, psbM
PK 169-0.1		PLNNB	AUR	05 19.2	+38 11	12	32 s			16.2	
PK 173-5.1	K2-1; SS 38	PLNNB	AUR	05 08.1	+30 48	12	132 s		3	18.2	pF, pL, R, nBM at 165X, 3* invol
B 12		DRKNB	CAM	04 30.0	+54 17	none	24 m		5Ir		Lies south following B9 complex

Designation	Alt. Designation	Type	Const	RA	Dec				Mag	Class	Remarks
B 8, 9, 11, 13		DRKNB	CAM	04 19.0	+55 03	none	150m			5lr	3 deg north of Cl NGC 1528
IC 3568	PK 123+34.1	PLNNB	CAM	12 33.1	+82 34	11.6	18s		12.9	2(2a)	planetary or neb *9.5, *13 p 15"
NGC 1501	PK 144+6.1	PLNNB	CAM	04 07.0	+60 55	12	56s	48s	14.4	3	pB, pS, vlE, 1'Diam
PK 118+2.1	Sh1-118	PLNNB	CAM	00 07.6	+64 58	12.9	120s			3	Typical spectrum of HII region
IC 2448	PK 285-14.1	PLNNB	CAR	09 07.1	-69 57	11.5	8s		14.2	2b	vS, R, nearly stellar; ring shape
IC 2501	PK 281-5.1	PLNNB	CAR	09 38.8	-60 06	11.3	2s		14.5	1	planetary, stellar
IC 2553	PK 285-5.1	PLNNB	CAR	10 09.3	-62 37	13	4s		15.5		planetary, stellar; Br PLNNB in fine field–Hartung
IC 2621	PK 291-4.1	PLNNB	CAR	11 00.3	-65 15	10.5	5s		15.4	1	planetary, stellar, 10.5 mag
NGC 2867	PK 278-5.1	PLNNB	CAR	09 21.4	-58 19	9.7	12.0s		16	4	!! = *8, vS, R, *15 np (90deg), *nr 13"
NGC 3211	PK 286-4.1	PLNNB	CAR	10 17.8	-62 40	11.8	12s		17.2	2b	plan = *10, R, am 150 st
NGC 3324	IC 2599	CL+NB	CAR	10 37.3	-58 40	6.7	16m			I 3 rn	pB, vvL, iF, D* inv
NGC 3372	Dunlop 309	BRTNB	CAR	10 45.1	-59 52	3	120m	120m		E	! great neb, Eta Carinae with Keyhole Nebula

!	remarkable object	!!	very remarkable object	E	elongated	s	suddenly
am	among	n	north	e	extremely	s	south
att	attached	N	nucleus	er	easily resolved	sc	scattered
bet	between	neb	nebula, nebulosity	F	faint	susp	suspected
B	bright	P w	paired with	f	following	st	star or stellar
b	brighter	p	pretty (before F, B, L, S)	g	gradually	v	very
C	compressed	p	preceding	iF	irregular figure	var	variable
c	considerably	P	poor	inv	involved	nf	north following
Cl	cluster	R	round	irr	irregular	np	north preceding
D	double	Ri	rich	L	large	sf	south following
def	defined	r	not well resolved	l	little	sp	south preceding
deg	degrees	rr	partially resolved	mag	magnitude	11m	11th magnitude
diam	diameter	rrr	well resolved	M	middle	8...	8th mag and fainter
dif	diffuse	S	small	m	much	9...13	9th to 13th magnitude

Object	Other	Type	Con	R.A.	Dec.	Mag	SIZE_MAX	SIZE_MIN	Class	BRSTR	Description and notes
PK 264-12.1	He2-5	PLNNB	CAR	07 47.4	-51 16	12.3	<101 s				
PK 278-4.1	He2-32	PLNNB	CAR	09 30.9	-57 36	12.4	40 s				
PK 279-3.1	He2-36	PLNNB	CAR	09 43.5	-57 17	10.4	<25 s			11.5	
PK 283-1.1	Hoffleit 4	PLNNB	CAR	10 15.6	-58 51	11.8	30 s		4		Ring shape
PK 283-4.1	He2-39	PLNNB	CAR	10 03.9	-60 45	12.8	10 s				Ring shape
PK 288+0.1	Hoffleit 38	PLNNB	CAR	10 54.6	-59 10	12.4	30 s		4		Ring shape
PK 289-0.1	He2-58	BRTNB	CAR	10 56.2	-60 27	11	35 s		4	8.5	Not a PLNNB; AG CAR variable W; Ring shaped faint neby
PK 290-0.1	Hoffleit 48	PLNNB	CAR	11 03.9	-60 36	12.6	20 s		3		
IC 289	PK 138+2.1	PLNNB	CAS	03 10.3	+61 19	12	45 s	30 s	4(2)	16.8	pB, pL, R, bet 2 vF stars
IC 1747	PK 130+1.1	PLNNB	CAS	01 57.6	+63 19	12	13 s		3b	15.8	Stellar
IC 1805	OCL 352	CL+NB	CAS	02 32.7	+61 27	6.5	60 m	60 m	III 3 p n	7.9	Cl, C, eL neby extends following; cluster is Mel 15
IC 1848	OCL 364	CL+NB	CAS	02 51.4	+60 25	6.5	40 m	10 m	IV 3 p n	7.1	Cl, star following, in F neby eL 90'X45'
NGC 281	IC 11	CL+NB	CAS	00 53.0	+56 37	7.4	4.0 m		E+*	9	F, vL, dif, S triple * on np edge
NGC 896	SG 1.04	BRTNB	CAS	02 25.5	+62 01	none	27 m	13 m	E		eF, pL, iF
NGC 7635	LBN 549	BRTNB	CAS	23 20.2	+61 11	11	15 m	8 m	E		vF, *8 inv; Bubble Nebula, L neb ring
PK 114-4.1	Abell 82	PLNNB	CAS	23 45.8	+57 04	12.7	94 s		3b	13	F, pS, R, nBM at 135X, visual size = 10 arcsec
PK 118-8.1	VY 1-1	PLNNB	CAS	00 18.7	+53 53	12.5	5 s		5Ir		
Be 146		DRKNB	CEN	13 57.6	-40 00	none	20 m	8 m			Adjacent to bright nebula NGC 5367
IC 2944	Cr 249	CL+NB	CEN	11 37.9	-63 21	4.5	60 m	35 m	II 1 p n	6.4	* 3.4 mag in vL neby; Lambda Cen cluster; neby vL, F

Name	Type	Const	RA	Dec	mag	size	size2	class	mag2	Remarks
NGC 3918	PLNNB	CEN	11 50.3	-57 11	8.4	12 s		2b	15.5	l, S, R, blue, = *7, d = 1'.5
NGC 5307	PLNNB	CEN	13 51.1	-51 12	12.1	15 s		3	14.7	Pln or vf, eS, Dneb
PK 290+7.1	PLNNB	CEN	11 28.6	-52 56	11.4	30 s	10 s	3(4)		
PK 292+4.1 =B 8	PLNNB	CEN	11 33.4	-57 06	12.8	5 s				
PK 293+1.1 =He2-70	PLNNB	CEN	11 35.2	-60 17	12	45 s	27 s			
PK 296-3.1 =He2-73	PLNNB	CEN	11 48.6	-65 08	12.9	<5 s				
PK 315-0.1 =He2-111	PLNNB	CEN	14 33.3	-60 50	13	30 s				
PK 316+8.1 =He2-108	PLNNB	CEN	14 18.2	-52 11	10.1	12 s				
B148-9	DRKNB	CEP	20 49.1	+59 32	none	3 m		5C		Two small clouds south of B150
B150	DRKNB	CEP	20 50.6	+60 18	none	60 m	3 m	5lr		Curved filament 1.6deg south of Eta Cep
B152	DRKNB	CEP	21 14.5	+61 45	none	15 m	3 m	5lr		Double cloud 1 deg south preceding Alpha Cep
B160	DRKNB	CEP	21 38.0	+56 14	none	30 m	15 m	4lr		Lies south of BRTNB IC 1396
B161	DRKNB	CEP	21 40.3	+57 49	none	13 m	3 m	6CoG		In north part of BRTNB IC 1396

!	remarkable object	!!	very remarkable object	E	elongated	s	suddenly

!	remarkable object	!!	very remarkable object
am	among	n	north
att	attached	N	nucleus
bet	between	neb	nebula, nebulosity
B	bright	P w	paired with
b	brighter	p	pretty (before F, B, L, S)
C	compressed	p	preceding
c	considerably	P	poor
Cl	cluster	R	round
D	double	Ri	rich
def	defined	r	not well resolved
deg	degrees	rr	partially resolved
diam	diameter	rrr	well resolved
dif	diffuse	S	small

E	elongated	s	suddenly
e	extremely	s	south
er	easily resolved	sc	scattered
F	faint	susp	suspected
f	following	st	star or stellar
g	gradually	v	very
iF	irregular figure	var	variable
inv	involved	nf	north following
irr	irregular	np	north preceding
L	large	sf	south following
l	little	sp	south preceding
mag	magnitude	11m	11th magnitude
M	middle	8 . . .	8th mag and fainter
m	much	9 . . . 13	9th to 13th magnitude

Object	Other	Type	Con	R.A.	Dec.	Mag	SIZE_MAX	SIZE_MIN	Class	BRSTR	Description and notes
B162	LDN 1095	DRKNB	CEP	21 41.1	+56 19	none	13 m	2 m	4Ir		Curved strip following B160
B163	LDN 1106	DRKNB	CEP	21 42.2	+56 42	none	4 m		4IrG		In a part of bright nebula IC 1396
B169-71	LDN 1151	DRKNB	CEP	21 58.9	+58 45	none	80 m		5Ir		Narrow curved lanes 3 deg north following IC 1396
B173-4	LDN 1164	DRKNB	CEP	22 07.4	+59 10	none	40 m		6Ir		Patchy lane north following B169
B365	LDN 1090	DRKNB	CEP	21 34.9	+56 43	none	22 m	3 m	4S		Lies south preceding bright nebula IC 1396
B367	LDN 1113	DRKNB	CEP	21 44.4	+57 12	none	3 m		5IrG		In preceding part of bright nebula IC 1396
IC 1396	OCL 222	CL+NB	CEP	21 39.1	+57 30	3.5	89 m		II 3 mn	3.8	F, eL neby, incl Struve 2816; Neby is 165X135'; cluster is Tr 37
NGC 40	PK 120+9.1	PLNNB	CEP	00 13.0	+72 31	10.7	60 s	40 s	3b(3)	11.5	F, vS, R, vsmbM, L*cont f; Lord Rosse saw spiral structure
NGC 7023	OCL 235	CL+NB	CEP	21 00.5	+68 10	7.1	5.0 m		E+*		*7 in eF, eL, neby; complex structure of bright and dark filaments
NGC 7129	OCL 240	CL+NB	CEP	21 43.0	+66 07	11.5	2.7 m		IV 2 p n:b	11.5	I, cF, pL, gbM***
NGC 7354	PK 107+2.1	PLNNB	CEP	22 40.3	+61 17	12.9	22 s	18 s	4(3b)	16.5	B, S, R, pgvlbM
PK 116+8.1	M2-55	PLNNB	CEP	23 31.9	+70 23	12.2	42 s	36 s	3	21	
NGC 246	PK 118-74.1	PLNNB	CET	00 47.1	-11 52	8.5	240 s	210 s	3b	10.9	
Be 142		DRKNB	CHA	11 09.5	-77 16	none	100 m		6		Near reflection nebula IC 2631
NGC 3195	PK 296-20.1	PLNNB	CHA	10 09.4	-80 52	11.5	40 s	30 s	3	17.8	! pB, S, lE, 3 S * nr

Be 145		DRKNB	CIR	14 48.6	-65 15	none	12 m	5 m	5		Near reflection nebula vdBH 63
NGC 5315	PK 309-4.2	PLNNB	CIR	13 54.0	-66 31	13	5 m		2	11.3	stellar = 10.5 mag
PK 318-2.1	He2-114	PLNNB	CIR	15 04.1	-60 53	11.1	30s	24s			Bipolar shape
PK 318-2.2	He2-116	PLNNB	CIR	15 06.0	-61 22	10.6	45s				
IC 2165	PK 221-12.1	PLNNB	CMA	06 21.7	-12 59	12.5	9s	7 s	3b	15.1	vS
NGC 2296	IC 452	BRTNB	CMA	06 48.7	-16 54	13	0.6 m	0.4 m	R		vF, vS, R; Near Sirius
NGC 2359	LBN 1041	BRTNB	CMA	07 18.5	-13 14	none	10 m	5 m	E	11	!!, vF, vvL, viF; Duck Neb; curved filam; cent* Wolf Rayet; UHC filter helps
PK 232-1.1	M1-13	PLNNB	CMA	07 21.2	-18 08	12.6	10s	9 s	?(6)		
PK 242-11.1	M3-1	PLNNB	CMA	07 02.8	-31 35	12.2	14s	7 m	E	14.1	
Sh2-301	Gum 5; RCW 6	BRTNB	CMA	07 09.8	-18 29	none	8 m		3a		
PK 219+31.1	Abell 31	PLNNB	CNC	08 54.2	+08 55	12.2	16.8 m	15.6 m	6lr	15.5	
Be 157		DRKNB	CRA	19 02.9	-37 08	none	55 m	18 m			Lies south following bright nebulae NGC 6726-7
IC 1297	PK 358-21.1	PLNNB	CRA	19 17.4	-39 37	11.5	8 s	6 s	E	12.9	Stellar, gaseous spectrum
NGC 6726	ESO 396-N13	BRTNB	CRA	19 01.7	-36 53	none	2 m	2 m	E		*6, 7 in F, pL neb; Complex neb region w var* TY & R CrA invl

!	remarkable object	!!	very remarkable object	E	elongated
am	among	n	north	e	extremely
att	attached	N	nucleus	er	easily resolved
bet	between	neb	nebula, nebulosity	F	faint
B	bright	P w	paired with	f	following
b	brighter	p	pretty (before F, B, L, S)	g	gradually
C	compressed	p	preceding	iF	irregular figure
c	considerably	P	poor	inv	involved
Cl	cluster	R	round	irr	irregular
D	double	Ri	rich	L	large
def	defined	r	not well resolved	l	little
deg	degrees	rr	partially resolved	mag	magnitude
diam	diameter	rrr	well resolved	M	middle
dif	diffuse	S	small	m	much

s	suddenly
s	south
sc	scattered
susp	suspected
st	star or stellar
v	very
var	variable
nf	north following
np	north preceding
sf	south following
sp	south preceding
11m	11th magnitude
8 ...	8th mag and fainter
9 ... 13	9th to 13th magnitude

Object	Other	Type	Con	R.A.	Dec.	Mag	SIZE_MAX	SIZE_MIN	Class	BRSTR	Description and notes
NGC 6729	ESO 396-N*15	BRTNB	CRA	19 01.9	-36 57	none	25 m	20 m	E+R		Var*(11 . . .) w neb
PK 352-7.1	Fleming 3;	PLNNB	CRA	18 00.2	-38 50	11.4	2 s		1		
Coalsack	V V 133	DRKNB	CRU	12 53.0	-63 00	none	400 m		3? Ir		Adjacent to Southern Cross
PK 298-0.1	He2-77	BRTNB	CRU	12 09.0	-63 16	11	13 s	8 s			Not a PLNNB; compact H II region
PK 298-1.2	He2-76	PLNNB	CRU	12 08.4	-64 12	12.4	16 s				
PK 299+2.1		PLNNB	CRU	12 23.8	-60 14	12.7	30 s				
PK 300-0.1	He2-84	PLNNB	CRU	12 28.8	-63 44	11.7	34 s	23 s			
PK 300+0.1	He2-83	PLNNB	CRU	12 28.7	-62 06	12.9	6 s	5 s			
NGC 4361	ESO 573-PN19	PLNNB	CRV	12 24.5	-18 47	10.9	80 s		3a(2)	13.2	vB, L, R, vsmbMN, r
B144	LDN 857	DRKNB	CYG	19 59.0	+35 00	none	360 m		1 Ir		Fish on the platter; preceding cluster NGC 6883
B145	LDN 865	DRKNB	CYG	20 02.8	+37 40	none	35 m	6 m	4 G		Triangular; north of B144
B146	LDN 860	DRKNB	CYG	20 03.5	+36 02	none	1 m		6		In B144 adjacent to BD +35 3930
B157	LDN 1075	DRKNB	CYG	21 33.7	+54 40	none	4 m		4 C G		8' preceding 8th-mag. star BD +54 2576
B164	LDN 1070	DRKNB	CYG	21 46.5	+51 04	none	12 m	6 m	5 K G		0.8 deg following Pi 1 Cyg
B168		DRKNB	CYG	21 53.2	+47 12	none	100 m		5 Ir		Narrow E-W lane; Cocoon nebula at f end
B343	LDN 880	DRKNB	CYG	20 13.5	+40 16	none	10 m	5 m	5 Ir G		1.7 deg preceeding Gamma Cyg
B346		DRKNB	CYG	20 26.7	+43 45	none	10 m	4 m	6 K		In patchy area 2.8 deg preced and 1.5 deg south of Deneb
B347		DRKNB	CYG	20 28.4	+39 55	none	10 m	1 m	5		Narrow streak 1.2 deg foll and 20' south of Gamma Cyg
B350		DRKNB	CYG	20 49.1	+45 53	none	3 m		6 C		14' south of Cyg

B352	DRKNB	CYG	20 57.1	+45 22	none	20 m		5Ir		Lies north of North America nebula
B361	DRKNB	CYG	21 12.9	+47 22	none	17 m		4CG		Has faint preceding extension
B362	DRKNB	CYG	21 24.0	+50 10	none	15 m	8 m	5EG		Adjacent to 9th mag star
B364	DRKNB	CYG	21 33.6	+54 33	none	40 m		5Ir		Narrow lanes of B157
IC 1318	BRTNB	CYG	20 27.9	+40 00	14.9	45 m	20 m	E		Gamma Cygni Neb, L patches of F nebulosity
IC 5070	BRTNB	CYG	20 50.8	+44 21	8	60 m	50 m	E		F, dif; Pelican Nebula, 56Cyg involved
IC 5146	CL+NB	CYG	21 53.4	+47 16	10	20 m	10 m	IV 2 p n	9.6	F, L, iR,*9.5 m invl, br + drk masses invl; Cocoon Nebula
M1-92	BRTNB	CYG	19 36.3	+29 33	11.7	8 s	16 s	R		pB, vS, nBM, sE in PAO at 220X; Footprint Nebula
NGC 6826	PLNNB	CYG	19 44.8	+50 32	8.8	27 s	24 s	3a(2)	10.7	Planetary, B, pL, R, *11 m; Blinking Planetary
NGC 6857	BRTNB	CYG	20 02.8	+33 31	11.4	0.8 m	5.0 s	E	14.3	F, am Milky Way st
NGC 6884	PLNNB	CYG	20 10.4	+46 28	12.6	5.6 s		2b	16.7	stellar
NGC 6888	BRTNB	CYG	20 12.8	+38 19	10	20 m	10 m	E		F, vL, vmE, ** att; Crescent Neb; Wolf-Rayet * invl

!	remarkable object	!!	very remarkable object	s	suddenly
am	among	n	north	s	south
att	attached	N	nucleus	sc	scattered
bet	between	neb	nebula, nebulosity	susp	suspected
B	bright	P w	paired with	st	star or stellar
b	brighter	p	pretty (before F, B, L, S)	v	very
C	compressed	p	preceding	var	variable
c	considerably	P	poor	nf	north following
Cl	cluster	R	round	np	north preceding
D	double	Ri	rich	sf	south following
def	defined	r	not well resolved	sp	south preceding
deg	degrees	rr	partially resolved	11m	11th magnitude
diam	diameter	rrr	well resolved	8 ...	8th mag and fainter
dif	diffuse	S	small	9 ... 13	9th to 13th magnitude
		E	elongated		
		e	extremely		
		er	easily resolved		
		F	faint		
		f	following		
		g	gradually		
		iF	irregular figure		
		inv	involved		
		irr	irregular		
		L	large		
		l	little		
		mag	magnitude		
		M	middle		
		m	much		

Object	Other	Type	Con	R.A.	Dec.	Mag	SIZE_MAX	SIZE_MIN	Class	BRSTR	Description and notes
NGC 6960	LBN 191	SNREM	CYG	20 45.7	+30 43	7	210 m	160 m			pB, cL, eiF, 52 Cyg invl; Veil Nebula western part
NGC 6992	CED 182B	SNREM	CYG	20 56.4	+31 43	7	60 m	8 m	R		eF, eL, eE, eiF; Veil Nebula eastern part
NGC 6995	CED 182C	SNREM	CYG	20 57.1	+31 13	7	12 m		R		F, eL, neb&st in groups
NGC 7000	LBN 373	BRTNB	CYG	21 01.8	+44 12	4	120 m	30 m	E		F, eeL, dif nebulosity; North America Nebula; Cl NGC 6997 invl
NGC 7008	PK 93+5.2	PLNNB	CYG	21 00.5	+54 33	12	86 s	69 s	3	13.9	cB, L, E45, r, ** att
NGC 7026	PK 89+0.1	PLNNB	CYG	21 06.3	+47 51	12	25 s	9 s	3a	14	pB, biN
NGC 7027	PK 84-3.1	PLNNB	CYG	21 07.0	+42 14	9.6	18 s	11 s	3a	16	eB, S
NGC 7048	PK 88-1.1	PLNNB	CYG	21 14.3	+46 17	11	60 s	50 s	3b	18	pF, pL, dif, iR, vlbM
PK 64+5.1	BD+30 3639	PLNNB	CYG	19 34.8	+30 31	9.6	5 s		4	10	vF, S; Campbell's hydrogen star, BD means Bonner Durchmusterung
PK 77+14.1	Abell 61	PLNNB	CYG	19 19.2	+46 15	13	200 s		2b	17.4	
PK 79+5.1	M4-17	PLNNB	CYG	20 09.0	+43 44	12.3	23 s	21 s	4(2)		
PK 86-8.1	Hu 1-2	PLNNB	CYG	21 33.1	+39 38	12.7	10 s	7 s	2	13.3	
NGC 6891	PK 54-12.1	PLNNB	DEL	20 15.1	+12 42	10.5	15.5 s	7 s	2a(2b)	12.3	stellar = 9.5 m
NGC 6905	PK 61-9.1; H IV 16	PLNNB	DEL	20 22.4	+20 06	12	44 s	38 s	3(3)	14	B, pS, R, 4S* nr; Blue Flash Nebula
NGC 1714	ESO 85-EN8	LMCDN	DOR	04 52.1	-66 56	none			E+*		vB, S, E or biN, bM, sp 2
NGC 1769	ESO 85-EN23	LMCCN	DOR	04 57.7	-66 28	none			E		B, L, iR, vsmbM
NGC 1770	IC 2117	LMCCN	DOR	04 57.3	-68 25	9			E+*		Cl+ neb, pL, pRi, *11 18
NGC 1814	ESO 85-SC36	LMCCN	DOR	05 03.8	-67 18	9			E+*		vF, R, s of 2 in Cl
NGC 1816	ESO 85-SC37	LMCCN	DOR	05 03.8	-67 16	9	16 m		III 1 m	11.2	vF, R, 2nd neb in Cl
NGC 1829	ESO 56-SC57	LMCCN	DOR	05 05.0	-68 03	8.5			E+*		F, pL, R, r
NGC 1850	ESO 56-SC70	LMCCN	DOR	05 08.7	-68 46	9.3	3.4 m		E+*		vB, L, lE, vmCM, rr, !
NGC 1874	ESO 56-EN84	LMCCN	DOR	05 13.2	-69 23	9			E+*		neb and Cl, biN

Name	Alt name	Type	Con	RA	Dec	Mag	Size	Size	Class	Description
NGC 1876		LMCCN	DOR	05 13.3	−69 22	9			E+*	pB, iR, biN, 2nd in group
NGC 1955	ESO 56-SC121	LMCCN	DOR	05 26.2	−67 30	9			E+*	Cl, Ri, 2nd of several
NGC 1962	ESO 56-SC122	LMCCN	DOR	05 26.3	−68 50	8.5	13 m	12 m	E	vF, pL, R, 1st of 4
NGC 1965	ESO 56-SC123	LMCCN	DOR	05 26.5	−68 48	8.5				F, S, 2nf of 4
NGC 1966	ESO 56-SC125	LMCCN	DOR	05 26.8	−68 49	8.5	13 m			pB, R, pslbM, 3rd of 4, in pL, irr Cl
NGC 1968	ESO 56-SC130	LMCCN	DOR	05 27.4	−67 28	9	20 m	20 m	E+*	Cl, Ri, 3rd of several
NGC 1970	ESO 56-SC127	LMCCN	DOR	05 26.9	−68 50	8.5				4th of 4
NGC 1974	NGC 1991	LMCCN	DOR	05 28.0	−67 25	9	9 m		E+*	Cl, L, irR
NGC 1983	ESO 56-SC133	LMCCN	DOR	05 27.7	−68 59	8.5				Cl, vL, pRi, iF
NGC 1991	NGC 1974	Cl+NB	DOR	05 28.0	−67 25	9	9 m		E+*	Cl, 4th of sev
NGC 2011	ESO 56-SC144	LMCCN	DOR	05 32.3	−67 31	9.5				vB, S, R, psmbM
NGC 2014	ESO 56-SC146	LMCCN	DOR	05 32.3	−67 41	8.5	20 m		E+*	Cl, pL, pC, iF, st9 … 15
NGC 2070	ESO 57-EN6	LMCCN	DOR	05 38.6	−69 06	8.3			E	!!! vB, vL, looped; 30 Dor cluster in LMC
NGC 2074	ESO 57-EN8	LMCCN	DOR	05 39.1	−69 30	8.5			E	pB, pL, mE, 5* inv
NGC 2100	ESO 57-SC25	LMCCN	DOR	05 42.2	−69 13	9.6	2.3 m	16 s		B, pL, irr R, rr
NGC 6543	PK 96+29.1	PLNNB	DRA	17 58.6	+66 38	8.3	22 s	11.3	3a(2)	vB, pS, sbMvSN; Cat's Eye Nebula
IC 2118	NGC 1909	BRTNB	ERI	05 04.5	−07 16	none	180 m	60 m	R	F, eL, iF, NGC 1779 inv s; Witchhead Nebula

!	remarkable object	!!	very remarkable object	E	elongated
am	among	n	north	e	extremely
att	attached	N	nucleus	er	easily resolved
bet	between	neb	nebula, nebulosity	F	faint
B	bright	P w	paired with	f	following
b	brighter	p	pretty (before F, B, L, S)	g	gradually
C	compressed	p	preceding	iF	irregular figure
c	considerably	P	poor	inv	involved
Cl	cluster	R	round	irr	irregular
D	double	Ri	rich	L	large
def	defined	r	not well resolved	l	little
deg	degrees	rr	partially resolved	mag	magnitude
diam	diameter	rrr	well resolved	M	middle
dif	diffuse	S	small	m	much

s	suddenly
s	south
sc	scattered
susp	suspected
st	star or stellar
v	very
var	variable
nf	north following
np	north preceding
sf	south following
sp	south preceding
11m	11th magnitude
8 … 13	8th mag and fainter
9 … 13	9th to 13th magnitude

Object	Other	Type	Con	R.A.	Dec.	Mag	SIZE_MAX	SIZE_MIN	Class	BRSTR	Description and notes
NGC 1535	PK 206-40.1	PLNNB	ERI	04 14.3	−12 44	10.4	20 s	17 s	4(2c)	12.1	vB, S, R, pS, vsbM, r
NGC 1360	PK 220-53.1	PLNNB	FOR	03 33.2	−25 52	9.4	360 s	270 s	3	11.3	*8in B, L neb, E ns
IC 443	LBN 844	SNREM	GEM	06 17.8	+22 49	12	50 m	40 m			F, narrow curved
NGC 2371	NGC 2372	PLNNB	GEM	07 25.6	+29 29	13	74 s	54 s	3a+6	14.8	B, S, R, bMN, p of Dneb
NGC 2372	NGC 2371	PLNNB	GEM	07 25.6	+29 30	13	74 s	54 s	3a(4)	14.8	pB, S, R, bMN, f of Dneb
NGC 2392	PK 197+17.1	PLNNB	GEM	07 29.2	+20 55	8.6	47 s	43 s	3b(3b)	10.6	B, S, R, *9M, *8 nf 100"; Eskimo Nebula; several shells
PK 189+7.1	M1-7	PLNNB	GEM	06 37.4	+24 01	13	38 s	20 s	2	18.5	9 mag * 40" NW
PK 194+2.1	J 900	PLNNB	GEM	06 26.0	+17 47	12.4	12 s	10 s	3b(2)	15.3	vS, B
PK 205+14.1	Abell 21	PLNNB	GEM	07 29.0	+13 15	14.1	10 m	6 m		15.9	F, pL, E at 100X with UHC; crescent shape; Medusa Nebula, low surface brightness
IC 5148	PK 2-52.1	PLNNB	GRU	21 59.6	−39 23	11	120 s	10 s	4	16.2	pB, L, lE, * att, annular;
IC 4593	PK 25+40.1	PLNNB	HER	16 11.7	+12 04	11	12.5 s	20 s	2(2)	11.2	S, F, Stellar
NGC 6058	PK 64+48.1	PLNNB	HER	16 04.4	+40 41	13	25 s	13 s	3(2)	13.8	pF, vS, R, stellar
NGC 6210	PK 43+37.1	PLNNB	HER	16 44.5	+23 48	9.7	20 s		2(3b)	12.5	vB, vS, R, disc
PK 47+42.1	Abell 39	PLNNB	HER	16 27.5	+27 54	12.9	174 s		2c	15.8	
PK 51+9.1	Hu 2-1	PLNNB	HER	18 49.8	+20 50	11.6	3 s			13.3	
PK 53+24.1	VY 1-2	PLNNB	HER	17 54.4	+28 00	12	5.2 s	4.1 s	2	17.6	
NGC 2610	PK 239+13.1	PLNNB	HYA	08 33.4	−16 09	13	50 s	47 s	4(2)	15.9	F, S, att to *13, *7 nf
NGC 3242	PK 261+32.1	PLNNB	HYA	10 24.8	−18 39	8.6	40 s	35 s	4(3b)	12.3	! vB, lE 147, 45"d, blue; Ghost of Jupiter
PK 248+29.1	Abell 34	PLNNB	HYA	09 45.6	−13 10	12.9	281 s	268 s	2b	16.3	
PK 283+25.1	K1-22	PLNNB	HYA	11 26.7	−34 22	12.1	188 s	174 s		16.9	
PK 303+40.1	Abell 35	PLNNB	HYA	12 53.6	−22 52	12	938 s	636 s	3a		
NGC 602	ESO 29-SC43	SMCCN	HYI	01 29.4	−73 33	none	1.5 m	0.7 m	E+*		
IC 5217	PK 100-5.1	PLNNB	LAC	22 23.9	+50 58	12.6	7.5 s	6 s	2	14.6	B, S, R, psbM*, r: two parts divided by rift
PK 100-8.1	Merrill 2-2	PLNNB	LAC	22 31.7	+47 48	11.9			1		stellar

Object	Designation	Type	Con	RA	Dec	Mag	Maj	Min	Class	*mag	Notes
IC 418	PK 215-24.1	PLNNB	LEP	05 27.5	-12 42	10.7	14s	11s	4	10.2	vS, B; 11th mag star in nebulous disk
PK 342+27.1	Merrill 2-1	PLNNB	LIB	15 22.3	-23 38	11.6	7s		2	18.4	pB, vS, R, nBM at 220X
B228		DRKNB	LUP	15 45.5	-34 24	none	240m		6Ir		Contains small bright reflection nebula
IC 4406	PK 319+15.1	PLNNB	LUP	14 22.4	-44 09	11	100s	37s	4(3)	17.4	E 80deg; pB diffuse disk-Hartung
NGC 5873	PK 331+16.1	PLNNB	LUP	15 12.8	-38 08	12	3s		2	15.5	stellar = 9.5 mag
NGC 5882	PK 327+10.1	PLNNB	LUP	15 16.8	-45 39	10.5	7s			13.4	vS, R, quite sharp
NGC 6026	PK 341+13.1	PLNNB	LUP	16 01.4	-34 33	12.5	54s	36s	4	13.3	F, S, R, gpmbM, tri* np
PK 327+13.1	He2-118	PLNNB	LUP	15 06.2	-43 01	12.7	<5s				vF, pL, nBM, sev* invol at 100X with UHC
PK 164+31.1		PLNNB	LYN	07 57.8	+53 25	14	400s		4	16	
NGC 6720	M 57	PLNNB	LYR	18 53.6	+33 02	9.4	86s	62s	4(3)	15.8	Ring neb, B, pL, cE; Ring Nebula, cent* variable 14th mag
NGC 6765	PK 62+9.1	PLNNB	LYR	19 11.1	+30 33	12.9	38s		5	16	F, S, E
NGC 1943	ESO 56-SC114	LMCCN	MEN	05 22.5	-70 09	12					pF, pS, iR, vglbM, *15 at 191, 60"
B 37-9		DRKNB	MON	06 32.8	+10 38	none	180m		5Ir		Near bright nebulae NGC 2245; 47

! remarkable object
am among
att attached
bet between
B bright
b brighter
C compressed
c considerably
Cl cluster
D double
def defined
deg degrees
diam diameter
dif diffuse

!! very remarkable object
n north
N nucleus
neb nebula, nebulosity
P w paired with
p pretty (before F, B, L, S)
p preceding
P poor
R round
Ri rich
r not well resolved
rr partially resolved
rrr well resolved
S small

E elongated
e extremely
er easily resolved
F faint
f following
g gradually
iF irregular figure
inv involved
irr irregular
L large
l little
mag magnitude
M middle
m much

s suddenly
s south
sc scattered
susp suspected
st star or stellar
v very
var variable
nf north following
np north preceding
sf south following
sp south preceding
11m 11th magnitude
8 . . . 8th mag and fainter
9 . . . 13 9th to 13th magnitude

Object	Other	Type	Con	R.A.	Dec.	Mag	SIZE_MAX	SIZE_MIN	Class	BRSTR	Description and notes
NGC 2237	OCL 511	CL+NB	MON	06 30.9	+05 03	5.5	80 m	60 m	E		pB, vvL, dif, part of eL nebs ring ar 2239; Rosette Neby
NGC 2238	LBN 948	BRTNB	MON	06 30.7	+05 01	6	80 m	60 m	E		F* in neby, part of eL nebs ring ar 2239; Rosette Neby
NGC 2244	NGC 2239	CL+NB	MON	06 31.9	+04 57	4.8	24.0 m		II 3 r n:b	5.8	Cl, beautiful, st sc (12 MON); Cl in Rosette Nebula
NGC 2245	LBN 904	BRTNB	MON	06 32.7	+10 09	none	2 m	2 m	R		pL, com, mbN sf alm *, *7-8 nf
NGC 2261	LBN 920	BRTNB	MON	06 39.2	+08 45	none	2 m	1 m	E+R		B, vmE 330 deg, N com = *11; Hubble's Variable Nebula, R Mon involved
NGC 2264	OCL 495	CL+NB	MON	06 41.0	+09 54	3.9	20.0 m		III 3 m n:	5	eL neb, 3 deg diam, densest 12' sp 15 MON
NGC 2346	PK 215+3.1	PLNNB	MON	07 09.4	-00 48	12.5	60 s	50 s	3b(6)	11.2	*10 att w S, vF, neb
IC 4191	PK 304-4.1	PLNNB	MUS	13 08.8	-67 39	12	18 s	11 s	2	16.4	Stellar
NGC 4071	PK 298-4.1	PLNNB	MUS	12 04.3	-67 19	12.9	75 s			19.2	vF, vS, R, bM*, am st
NGC 5189	PK 307-3.1	PLNNB	MUS	13 33.5	-65 58	10.3	140 s		5	14.5	B, pL, cE, bM, curved axis
PK 307-4.1		PLNNB	MUS	13 39.5	-67 23	12.9	16 s	10 s			
PK 307-9.1	He2-97	PLNNB	MUS	13 45.4	-71 29	12.6	<5 s				
Sandqvist 149	Dark Doodad	DRKNB	MUS	12 25.0	-72 00	none	180 m	12 m			Just north of Gamma Mus; loops toward globular NGC 4833
PK 322-2.1	Menzel 1	PLNNB	NOR	15 34.2	-59 09	12.5	50 s	50 s	4(6)		
PK 325-4.1	He2-141	PLNNB	NOR	15 59.1	-58 24	12.4	16 s			13.6	
PK 329-2.2	Menzel 2; V V 78	PLNNB	NOR	16 14.5	-54 57	11.9	25 s	12 s	4(3)		
PK 330+4.1	Canon 1-1	PLNNB	NOR	15 51.3	-48 45	12.9	21 s		1		

B 42, 44-7		DRKNB	OPH	16 38.0	−24 06	none	600 m		6 lr	Contains B51 and B238; Narrow dk lanes extending foll Rho Oph
B 46	LDN 1775	DRKNB	OPH	16 57.2	−22 44	none	12 m		6 lrG	30' north of 24 Oph
B 57	LDN 11	DRKNB	OPH	17 08.3	−22 50	none	5 m		6 EG	In patchy region following B44
B 59, 65-7	LDN 1773	DRKNB	OPH	17 21.0	−27 00	none	300 m		6 lr	2 deg south of Theta Oph; Stem of Pipe nebula
B 60, 246	LDN 17	DRKNB	OPH	17 11.8	−22 27	none	30 m	20 m	3	In patchy region following B44
B 61		DRKNB	OPH	17 15.2	−20 21	none	10 m	4 m	6 lr	1 deg south preceding B63
B 62	LDN 100	DRKNB	OPH	17 16.2	−20 53	none	25 m		6 lr	30' south preceding B63
B 63	LDN 99	DRKNB	OPH	17 16.0	−21 23	none	100 m	15 m	3 lrG	3 deg north-north foll Theta Oph; w a globule at prec end
B 64	LDN 173	DRKNB	OPH	17 17.2	−18 32	none	20 m		6 Co	30' preceding globular cluster M9
B 67a	LDN 102	DRKNB	OPH	17 22.5	−21 53	none	16 m		6 lrG	3 deg north of Theta Oph
B 68	LDN 57	DRKNB	OPH	17 22.6	−23 44	none	4 m		6 KG	20' south preceding B72
B 69	LDN 55	DRKNB	OPH	17 22.9	−23 53	none	4 m		6 lr	15' south following B68

!	very remarkable object	E	elongated	s	suddenly
!!	remarkable object	e	extremely	s	south
am	among	er	easily resolved	sc	scattered
att	attached	F	faint	susp	suspected
bet	between	f	following	st	star or stellar
B	bright	g	gradually	v	very
b	brighter	iF	irregular figure	var	variable
C	compressed	inv	involved	nf	north following
c	considerably	irr	irregular	np	north preceding
Cl	cluster	L	large	sf	south following
D	double	l	little	sp	south preceding
def	defined	mag	magnitude	11m	11th magnitude
deg	degrees	M	middle	8 . . .	8th mag and fainter
diam	diameter	m	much	9 . . . 13	9th to 13th magnitude
dif	diffuse	n	north		
		N	nucleus		
		neb	nebula, nebulosity		
		P w	paired with		
		p	pretty (before F, B, L, S)		
		p	preceding		
		P	poor		
		R	round		
		Ri	rich		
		r	not well resolved		
		rr	partially resolved		
		rrr	well resolved		
		S	small		

Object	Other	Type	Con	R.A.	Dec.	Mag	SIZE_MAX	SIZE_MIN	Class	BRSTR	Description and notes
B 70	LDN 54	DRKNB	OPH	17 23.5	−23 58	none	4 m		4C?		20' south following B68
B 72	LDN 66	DRKNB	OPH	17 23.5	−23 38	none	30 m		6 SG		The Snake; 1.5 deg nnf Theta Oph
B 74		DRKNB	OPH	17 25.2	−24 12	none	15 m	10 m	5 Ir		15' preceding 44 Oph
B 75, 261-2	LDN 91	DRKNB	OPH	17 25.3	−22 28	none	110 m		5 Ir		Two arcs 1 deg north following B72
B 77, 269	LDN 69	DRKNB	OPH	17 28.0	−23 22	none	100 m		3 Ir		Faint extension of bowl of Pipe nebula
B 78	LDN 42	DRKNB	OPH	17 33.0	−26 00	none	200 m		6 Ir		2.5 deg south following Theta OPH; Bowl of Pipe nebula
B 79, 276	LDN 219	DRKNB	OPH	17 39.5	−19 47	none	50 m	30 m	6 Ir		B79 is narrow; straight north preceding extension
B244	LDN 1736	DRKNB	OPH	17 10.1	−28 24	none	20 m	30 m	5 Ir		Lies south of tip of Pipe nebula
B256	LDN 1749	DRKNB	OPH	17 12.2	−28 51	none	50 m	10 m	5 Ir		1.5 deg south of stem of Pipe nebula; Curved
B259	LDN 177	DRKNB	OPH	17 22.0	−19 19	none	30 m		4 Ir		50' south following globular cluster M9
B268, 270	LDN 178	DRKNB	OPH	17 32.0	−20 32	none	120 m		5 Ir		
IC 4604	LBN 1112	BRTNB	OPH	16 25.6	−23 27	none	60 m	25 m	E+*		Rho Ophiuchi in eL, vF, Irr neby; photographically 140'X70'
IC 4634	PK 0+12.1	PLNNB	OPH	17 01.6	−21 50	12	20 s	10 s	2a(3)	17	eS, B; Pale blue disk 10" across—Hartung
NGC 6309	PK 9+14.1	PLNNB	OPH	17 14.1	−12 55	11.6	20 s	10 s	3b(6)	16.3	B, S, bet 2*v nr; Box Nebula
NGC 6369	PK 2+5.1; HIV 11	PLNNB	OPH	17 29.3	−23 46	11	30 s	29 s	4(2)	15.1	Annular, pB, S, R; Little Ghost Nebula
NGC 6572	PK 34+11.1	PLNNB	OPH	18 12.1	+06 51	8	15 s	12 s	2a	12	vB, vS, R, l hazy

PK 3+2.1	Hubble 4	PLNNB	OPH	17 41.9	−24 42	13	6.6s	5.8s	3b	14.9	pF, S, comet-shaped at 220X; * invol
PK 8+5.1	The 4-2	PLNNB	OPH	17 46.2	−18 40	13	20s				
PK 8+6.1	He2-260	PLNNB	OPH	17 38.9	−18 18	11	<10s				
PK 357+7.1	M4-3	PLNNB	OPH	17 10.7	−27 09	12.9	<10s				
B 30-2, 225		DRKNB	ORI	05 29.8	+12 32	none	80m	55m	5lr		3 deg north preceding Lamba Ori
B 33		DRKNB	ORI	05 40.9	−02 28	none	6m	4m	4lr		Horsehead nebula; part of large dark following cloud
B 35		DRKNB	ORI	05 45.5	+09 03	none	20m	10m	5 E		Near FU ORI and bright nebula Ced 59
B 36		DRKNB	ORI	05 45.7	+07 31	none	120m		4lr		Narrow south preceding-north following lane
IC 434	LBN 954	BRTNB	ORI	05 41.0	−02 27	11	90m	30m	E		eF, vvL, vmE; 1 deg long incl Zeta Ori; contains dark Horsehead nebula (B 33)
NGC 1973	CED 55B	BRTNB	ORI	05 35.1	−04 44	7	5 m	5 m	E		*8, 9 inv in neb
NGC 1975	CED 55C	BRTNB	ORI	05 35.3	−04 41	7	10m	5 m	E		B** inv in neb

Abbr.	Meaning	Abbr.	Meaning	Abbr.	Meaning
!!	very remarkable object	!	remarkable object	s	suddenly
am	among	n	north	s	south
att	attached	N	nucleus	sc	scattered
bet	between	neb	nebula, nebulosity	susp	suspected
B	bright	P w	paired with	st	star or stellar
b	brighter	p	pretty (before F, B, L, S)	v	very
C	compressed	p	preceding	var	variable
c	considerably	p	poor	nf	north following
Cl	cluster	R	round	np	north preceding
D	double	Ri	rich	sf	south following
def	defined	r	not well resolved	sp	south preceding
deg	degrees	rr	partially resolved	11m	11th magnitude
diam	diameter	rrr	well resolved	8 . . . 13	8th mag and fainter
dif	diffuse	S	small	9 . . . 13	9th to 13th magnitude
E	elongated				
e	extremely				
er	easily resolved				
F	faint				
f	following				
g	gradually				
iF	irregular figure				
inv	involved				
irr	irregular				
L	large				
l	little				
mag	magnitude				
M	middle				
m	much				

Object	Other	Type	Con	R.A.	Dec.	Mag	SIZE_MAX	SIZE_MIN	Class	BRSTR	Description and notes
NGC 1976	M 42	CL+NB	ORI	05 35.3	-05 23	4	90 m	60 m	E+R		!!!, Theta Orionis and the great nebula M42
NGC 1977	OCL 525	CL+NB	ORI	05 35.3	-04 51	7	20 m	10 m	E+R		42 Orionis neb
NGC 1980	OCL 529	CL+NB	ORI	05 35.4	-05 55	2.5	14 m	14 m	III 3 m n:		vF, vvL, 44 Ori invl
NGC 1982	M 43	BRTNB	ORI	05 35.5	-05 16	9	20 m	15 m	E		!, vB, vL, R w tail, mbM*8
NGC 2022	PK 196-10.1	PLNNB	ORI	05 42.1	+09 05	12.8	28 s	27 s	4(2)	15.8	pB, vS, vlE
NGC 2024	CED 55P	BRTNB	ORI	05 41.7	-01 51	none	30 m	30 m	E		! irr, B, vvL, black sp incl
NGC 2068	M 78	BRTNB	ORI	05 46.8	+00 05	8	8 m	6 m	E		B, L, wisp, gmbN, 3* inv, r; comet shaped
NGC 2071	LBN 938	BRTNB	ORI	05 47.1	+00 18	8	7 m	5 m	R		D*(10 & 14m) w vF, L chev
NGC 2175	OCL 476	CL+NB	ORI	06 09.6	+20 29	6.8	18.0 m		IV 3 p n	7.6	*8 in neb
PK 190-17.1	J 320	PLNNB	ORI	05 05.6	+10 42	12.9	11 s	8 s	2(4)	13.5	vF, S, R
PK 197-14.1	Abell 10	PLNNB	ORI	05 31.8	+06 56	12.7	34 s		3	19.5	
Sh2-264		BRTNB	ORI	05 35.0	+10 00	none	270 m		E		Seems to apply to entire Lambda Ori neb
Sh2-276		BRTNB	ORI	05 48.0	+01 00	none	600 m		E		Barnard's Loop, faint, long neby
PK 320-28.1	He2-434	PLNNB	PAV	19 33.8	-74 33	12.2	8 s	6 s	3b		vF, vL, irr; only seen with UHC filter 100X
PK 104-29.1	Jones 1	PLNNB	PEG	23 35.9	+30 28	12.7	314 s			16.2	
B 1, 2, 202-6		DRKNB	PER	03 32.1	+31 10	none	160 m		5 Ir		Patchy region south preceding NGC 1333
B 3, 4	LDN 1470	DRKNB	PER	03 44.0	+31 47	none	100 m		5 Ir		Lies south preceding O Per
B 5	LDN 1471	DRKNB	PER	03 48.0	+32 54	none	22 m	9 m	5 EG		1 deg north following O Per
IC 348	IC 1985	CL+NB	PER	03 44.6	+32 10	7.3	10 m	10 m	IV 2 p n	8.5	pB, vL, vgbM; surounds Omicron Persei
IC 351	PK 159-15.1	PLNNB	PER	03 47.6	+35 03	12.4	8 s	6 s	2a	15	plan neb = * 10 mag, 9 mag * p

Object	Desig.	Const.	RA	Dec	Mag	Size1	Size2	Type		Notes
IC 2003	PK 161-14.1	PER	03 56.4	+33 53	12.6	7 s		2	15.3	pB, eS, lE ns,*13 n 4″,*12 sp 18″
NGC 650	M 76	PER	01 42.3	+51 35	11	163 s	107 s	3[6]	17.6	vB, p of Dneb; Little Dumbbell Nebula
NGC 651	M 76	PER	01 42.3	+51 35	11	163 s	107 s	3[6]	17.6	vB, f of Dneb; Little Dumbbell Nebula
NGC 1333	LBN 741	PER	03 29.2	+31 22	none	9 m	7 m	R	9.5	F, L,*10 nf; Bright neb with dark neb including B 205
NGC 1491	LBN 704	PER	04 03.2	+51 19	none	9 m	6 m	E		vB, S, iF, bM, r, * inv
NGC 1499	LBN 756	PER	04 03.2	+36 22	5	160 m	40 m	E		vF, vL, mE ns, dif; California Nebula 0.6deg from Xi Per
NGC 1624	OCL 403	PER	04 40.6	+50 28	11.8	1.9 m	5 m	II 1 p n:b11.8		F, cL, iF, 6or7*+ neb
Be 135	DRKNB	PUP	07 19.0	-44 35	none	13 m		6E		Contains small reflection nebula
NGC 2438	PK 231+4.2	PUP	07 41.8	-14 44	11	65 s	20 s	4[2]	17.5	pB, pS, vlE, r, on edge of Cl M 46
NGC 2440	PK 234+2.1	PUP	07 41.9	-18 13	11.5	54 s		5[3]	17.5	cB, not v well def

!	remarkable object		E	elongated
am	among		e	extremely
att	attached		er	easily resolved
bet	between		F	faint
B	bright		f	following
b	brighter		g	gradually
C	compressed		iF	irregular figure
c	considerably		inv	involved
Cl	cluster		irr	irregular
D	double		L	large
def	defined		l	little
deg	degrees		mag	magnitude
diam	diameter		M	middle
dif	diffuse		m	much
!!	very remarkable object		s	suddenly
n	north		s	south
N	nucleus		sc	scattered
neb	nebula, nebulosity		susp	suspected
P w	paired with		st	star or stellar
p	pretty (before F, B, L, S)		v	very
p	preceding		var	variable
P	poor		nf	north following
R	round		np	north preceding
Ri	rich		sf	south following
r	not well resolved		sp	south preceding
rr	partially resolved		11m	11th magnitude
rrr	well resolved		8 . . .	8th mag and fainter
S	small		9 . . . 13	9th to 13th magnitude

Object	Other	Type	Con	R.A.	Dec.	Mag	SIZE_MAX	SIZE_MIN	Class	BRSTR	Description and notes
NGC 2452	PK 243-1.1	PLNNB	PUP	07 47.4	-27 20	12.6	22s	16s	4(3)	17.5	F, S, lE, am 60*
NGC 2467	OCL 668	CL+NB	PUP	07 52.5	-26 26	7.1	15.0m		l 3 m n:b		pB, vL, R, er,*8 m
NGC 2579	OCL 724	CL+NB	PUP	08 20.9	-36 13	7.5	10.0m		IV 2 p :b 9.5		D* in pS neb, am 70*
PK 248-8.1	M4-2	PLNNB	PUP	07 28.9	-35 45	13	8s		2		F, S, R, nBM; held steady at 220X
NGC 2818A	PK 261+8.1	PLNNB	PYX	09 16.0	-36 36	11.9	36s	36s	3b	16.1	l pB, pL, R, vglbM, in L Cl
PK 254+5.1	M3-6	PLNNB	PYX	08 40.7	-32 23	11	11s	6s	2a		
B 40		DRKNB	SCO	16 14.7	-18 59	none	15 m		3lr		In bright nebula 50' north following Nu Sco
B 41, 43		DRKNB	SCO	16 22.0	-19 40	none	200m		6E		Two clouds 2.5 deg and 4 degrees following Nu Sco
B 44a	SL 18	DRKNB	SCO	16 44.8	-40 23	none	5 m		5lr		2.4 degrees preceding bright nebula IC 4628
B 48	SL 20	DRKNB	SCO	17 01.0	-40 47	none	40 m	15 m	5lr		1 deg south following bright nebula IC 4628
B 50	SL 30	DRKNB	SCO	17 03.0	-34 26	none	15 m		6lr		30' south preceding star CoD -33 degrees 11706
B 53	SL 32	DRKNB	SCO	17 06.1	-33 15	none	30 m	10 m	4lr		Curved
B 55-6	LDN 1682	DRKNB	SCO	17 07.5	-23 00	none	30 m	10 m	5lr		
B 58	SL 23	DRKNB	SCO	17 11.2	-40 25	none	15 m		6lr		2.7 deg following bright nebula IC 4628
B231	SL 24	DRKNB	SCO	16 37.5	-35 12	none	50 m	40 m	6lr		
B233	SL 25	DRKNB	SCO	16 44.1	-35 21	none	55 m	20 m	5lr		1 deg following B231
B235	SL 15	DRKNB	SCO	16 46.6	-44 30	none	7 m	3 m	6E		
B252	LDN 1698	DRKNB	SCO	17 15.2	-32 13	none	20 m	5 m	5lr		
B257		DRKNB	SCO	17 22.0	-35 35	none	10 m	7 m	5		Faint reflection nebula at edge
B263	SL 22	DRKNB	SCO	17 26.3	-42 38	none	30 m		5lr		

Name	Alt name	Type	Const	RA	Dec	Mag	Maj	Min	Class	Mag*	Remarks
B283		DRKNB	SCO	17 51.3	-33 53	none	90 m	60 m	5lr		1 deg north-north preceding cluster M7
B287		DRKNB	SCO	17 54.4	-35 12	none	25 m	15 m	5lr		30' south-south following cluster M7
Be 149		DRKNB	SCO	16 09.4	-39 08	none	60 m	12 m	6lr		Contains faint reflection nebula
IC 4599	PK 321+5.1	PLNNB	SCO	16 19.4	-42 16	12.3	16 s	13 s		16.3	annular
IC 4628	ESO 332-EN14	BRTNB	SCO	16 57.0	-40 27	none	90 m	60 m	E		F, eL, E pf, dif; Table of Scorpius
IC 4663	PK 346-8.1	PLNNB	SCO	17 45.5	-44 54	13	14 s	12 s	4	14	vF, S, nearly stellar
NGC 6153	PK 341+5.1	PLNNB	SCO	16 31.5	-40 15	11.5	25 s		4	15.5	Planetary, stellar
NGC 6302	PK 349+1.1	PLNNB	SCO	17 13.7	-37 06	12.8	72 s	30 s	6	16.6	pB, E pf; Bug Nebula flattened figure 8 shape
NGC 6334	ESO 392-EN9	BRTNB	SCO	17 20.8	-36 06	none	120 m	110 m	E		cF, vL, lCF, vglBf, *8inv
NGC 6337	PK 349-1.1	PLNNB	SCO	17 22.3	-38 29	12.3	38 s	28 s	4	14.8	ring neb, eF, S, am st
NGC 6357	ESO 392-SC10	BRTNB	SCO	17 24.7	-34 12	none	50 m	40 m	E+*		F, L, E, vglbM, D* inv
NGC 6480	ESO 456-?13	CL+NB	SCO	17 54.4	-30 27	12	5 m		E+*		neb S part of Milky Way
PK 342-2.1	He2-198	PLNNB	SCO	17 06.4	-44 13	13	25 s	15 s	4		
PK 342-4.1	He2-207	PLNNB	SCO	17 19.5	-45 53	12	40 s	26 s			

! remarkable object	!! very remarkable object	E elongated	s suddenly
am among	n north	e extremely	s south
att attached	N nucleus	er easily resolved	sc scattered
bet between	neb nebula, nebulosity	F faint	susp suspected
B bright	P w paired with	f following	st star or stellar
b brighter	p pretty (before F, B, L, S)	g gradually	v very
C compressed	p preceding	iF irregular figure	var variable
c considerably	P poor	inv involved	nf north following
Cl cluster	R round	irr irregular	np north preceding
D double	Ri rich	L large	sf south following
def defined	r not well resolved	l little	sp south preceding
deg degrees	rr partially resolved	mag magnitude	11m 11th magnitude
diam diameter	rrr well resolved	M middle	8… 8th mag and fainter
dif diffuse	S small	m much	9…13 9th to 13th magnitude

Object	Other	Type	Con	R.A.	Dec.	Mag	SIZE_MAX	SIZE_MIN	Class	BRSTR	Description and notes
PK 344+4.1	Vd1-1	PLNNB	SCO	16 42.6	-38 55	12	<10 s		1		
PK 345-4.1	Canon 1-3	PLNNB	SCO	17 26.3	-44 12	11.9	<5 s				
PK 349-4.1	Longmore 16	PLNNB	SCO	17 35.7	-40 12	13	83 s		2	16	
PK 351+5.1	M2-5	PLNNB	SCO	17 02.3	-33 10	13	5.1 s				
PK 355+3.2	H1-9	PLNNB	SCO	17 21.5	-30 21	10	<10 s				
PK 356-4.1	Canon 2-1	PLNNB	SCO	17 54.6	-34 23	12.2	3 s	2 s		14	pB, vS, lE, central * seen at 220X; on N side of M7
B 95	LDN 406	DRKNB	SCT	18 25.6	-11 45	none	30 m		5CG		2.6deg north following bright nebula M16
B 97	LDN 435	DRKNB	SCT	18 29.1	-09 56	none	50 m	50 m	4lr		1 deg north preceding cluster NGC 6649
B100-1	LDN 443	DRKNB	SCT	18 32.7	-09 08	none	40 m	15 m	5lrg		1.4deg north of cluster NGC 6649 and curved
B103	LDN 497	DRKNB	SCT	18 39.2	-06 37	none	40 m	40 m	6lr		On north preceding side of Scutum star cloud
B104	LDN 532	DRKNB	SCT	18 47.3	-04 32	none	16 m	1 m	5		20' north of Beta Sct; L-shaped
B108		DRKNB	SCT	18 49.6	-06 19	none	3 m		3		0.5 deg preceding cluster M11
B110	LDN 530	DRKNB	SCT	18 50.2	-04 46	none	9 m		6lrG		Part of B111
B111, 119a	LDN 534	DRKNB	SCT	18 51.0	-05 00	none	120 m		3lr		Two crescent-shaped areas north of M11
B112		DRKNB	SCT	18 51.2	-06 40	none	20 m	20 m	4lr		Lies south of M11
B113		DRKNB	SCT	18 51.4	-04 19	none	11 m		5lrG		Part of B111
B114-8	LDN 548	DRKNB	SCT	18 53.2	-07 06	none	50 m	5 m	6lr		Chain of dark neb south following M11
B118	LDN 509	DRKNB	SCT	18 53.9	-07 27	none	1 m		6CG		Member of group
B312	LDN 379	DRKNB	SCT	18 30.9	-15 08	none	100 m		4E		2.5 deg following Omega nebula w sharp n boundary

Designation	Other name	Type	Const.	RA	Dec	Mag	Dim 1	Dim 2	Class	CS mag	Remarks
B314	LDN 445	DRKNB	SCT	18 37.7	−09 37	none	35 m	25 m	5lr		1 degree north following cluster NGC 6649
B318		DRKNB	SCT	18 49.7	−06 24	none	90 m	2 m	2		Narrow preceding-following lane just south of cluster M11
IC 1295	PK 25-4.2	PLNNB	SCT	18 54.6	−08 50	12.7	102 s	87 s	3b(2)	15	pL, eF, 0.4′ ESE from NGC 6712
PK 15-3.1	Abell 44	PLNNB	SCT	18 30.2	−16 45	12.6	63 s	39 s	2		
PK 19-4.1	M1-60	PLNNB	SCT	18 43.7	−13 45	12.3			1		
PK 19-5.1	M1-61	PLNNB	SCT	18 45.9	−14 28	12.5	<5 s		1		
PK 20-0.1	Abell 45	PLNNB	SCT	18 30.3	−11 37	12.9	302 s	281 s	3b	20.1	
PK 21-1.1	M1-51	PLNNB	SCT	18 33.5	−11 07	13	3.9 s	3.0 s	3		
PK 22-3.1	M1-58	PLNNB	SCT	18 43.0	−11 07	12.4	7.0 s	5.8 s	2		
PK 23-2.1	M1-59	PLNNB	SCT	18 43.4	−09 05	12.4	4.8 s	4.3 s	2		
NGC 6611	M 16	CL+NB	SER	18 18.8	−13 47	6	7 m		II 3 m n:a 11		L, B, scattered Cl, neb invl: Star Queen Nebula
PK 13+4.1	Shane 1	PLNNB	SER	17 59.0	−15 32	13	5.0 s	4.8 s	2		
PK 13+32.1		PLNNB	SER	16 21.1	+00 17	12.8	5 s			14.7	
PK 19+3.1	M3-25	PLNNB	SER	18 15.3	−10 10	12.5	4.3 s	3.5 s			
PK 32+7.2	PC 19	PLNNB	SER	18 24.7	+02 30	12.2					

l	remarkable object	E	elongated	s	suddenly	
am	among	e	extremely	s	south	
att	attached	er	easily resolved	sc	scattered	
bet	between	F	faint	susp	suspected	
B	bright	f	following	st	star or stellar	
b	brighter	g	gradually	v	very	
C	compressed	iF	irregular figure	var	variable	
c	considerably	inv	involved	nf	north following	
Cl	cluster	irr	irregular	np	north preceding	
D	double	L	large	sf	south following	
def	defined	l	little	sp	south preceding	
deg	degrees	mag	magnitude	11 m	11th magnitude	
diam	diameter	M	middle	8 . . . 13	8th mag and fainter	
dif	diffuse	m	much	9 . . . 13	9th to 13th magnitude	
ll	very remarkable object					
n	north					
N	nucleus					
neb	nebula, nebulosity					
P w	paired with					
p	pretty (before F, B, L, S)					
p	preceding					
P	poor					
R	round					
Ri	rich					
r	not well resolved					
rr	partially resolved					
rrr	well resolved					
S	small					

Object	Other	Type	Con	R.A.	Dec.	Mag	SIZE_MAX	SIZE_MIN	Class	BRSTR	Description and notes
IC 4997	PK 58-10.1	PLNNB	SGE	20 20.1	+16 44	11.3	2.0 s	1.4 s	1	13.7	Stellar
NGC 6879	PK 57-8.1	PLNNB	SGE	20 10.4	+16 55	11	4.7 s	4.1 s	2a	15	stellar = 10m
NGC 6886	PK 60-7.2	PLNNB	SGE	20 12.7	+19 59	12.5	4 s		2(3)	16.5	stellar = 10m
Sh2-82		BRTNB	SGE	19 30.3	+18 16	none	7 m	7 m	E+R		pF, pL, R, nBM, 2 * invl at 100X; UHC helps
Sh2-84		BRTNB	SGE	19 49.0	+18 24	none	15 m	3 m	E		pF, pL, mE, S side B at 135X with UHC filter
B 84	LDN 235	DRKNB	SGR	17 46.5	−20 11	none	30 m	15 m	6lr		1.5deg north; 40' following 58 Oph; B83a nearby
B 85		DRKNB	SGR	18 02.6	−23 02	none					Dark regions in Trifid nebula
B 86	LDN 93	DRKNB	SGR	18 02.7	−27 50	none	4 m		5lrG		On Sgr star cloud preceding cluster NGC 6520; Ink Spot Neb
B 87	LDN 1771	DRKNB	SGR	18 04.3	−32 30	none	12 m		4CG		Parrot's Head; 4.5deg south of cluster NGC 6520
B 88-9, 286		DRKNB	SGR	18 03.8	−24 23	none					Dark regions in Lagoon nebula
B 90	LDN 108	DRKNB	SGR	18 10.2	−28 19	none	10 m		6lrG		1.5deg following; 20' south of cluster NGC 6520
B 91	LDN 227	DRKNB	SGR	18 10.0	−23 39	none	5 m	2 m	5K		Adjacent to bright nebulae IC 1274-5
B 92	LDN 323	DRKNB	SGR	18 15.5	−18 11	none	12 m	6 m	6EG		On NW edge of Small Sagittarius Star Cloud 30' following B92
B 93	LDN 327	DRKNB	SGR	18 16.9	−18 04	none	12 m	2 m	4CoG		
B303	LDN 210	DRKNB	SGR	18 09.2	−24 07	none	1 m		5S		In bright nebula IC 4685
IC 4673	PK 3-2.3	PLNNB	SGR	18 03.3	−27 06	13	18 s	12.5 s	4	14.6	annular, 13 mag * nf 33"

Object	PK / other	Type	Const	RA	Dec	mag	size	size	Class	CS mag	Description
IC 4776	PK 2-13.1	PLNNB	SGR	18 45.8	-33 21	12.5	8 s	6 s	2a	16	vS, F
NGC 6439	PK 11+5.1	PLNNB	SGR	17 48.3	-16 28	13	6.1 s	5.1 s	2a	18	stellar = 13 m
NGC 6445	PK 8+3.1; H II 586	PLNNB	SGR	17 49.3	-20 01	13	35 s	30 s	3b(3)	19	pB, pS, R, gbM, r, *15 np
NGC 6514	M 20	Cl+NB	SGR	18 02.7	-22 58	6.3	28.0 m		E+*	6	vB, vL, Trifid, D* inv; Trifid Nebula
NGC 6523	M 8	Cl+NB	SGR	18 03.7	-24 23	5	45 m	30 m	E		!!!, vB, eL, eiF, w L Cl; Lagoon Nebula
NGC 6537	PK 10+0.1	PLNNB	SGR	18 05.2	-19 51	12	5 s	5 s	2a(6)	19.5	B, S, stellar
NGC 6563	PK 358-7.1	PLNNB	SGR	18 12.0	-33 52	13	54 s	41 s	3a	18	F, L, cE, hazy edge
NGC 6565	PK 3-4.5	PLNNB	SGR	18 11.9	-28 11	13	10 s	8 s	4	19.5	stellar
NGC 6567	PK 11-0.2	PLNNB	SGR	18 13.8	-19 05	11.5	11 s	7 s	2a(3)	15	stell, 11 m, in a Cl
NGC 6618	M 17	Cl+NB	SGR	18 20.8	-16 11	6	11.0 m		III 3 m n: 9.3		!!!, B, eL, eiF, 2 hooked; Swan Nebula
NGC 6629	PK 9-5.1; H II 204	PLNNB	SGR	18 25.7	-23 12	10.5	16 s	14 s	2a	12.9	pB, eeS, R
NGC 6644	PK 8-7.2	PLNNB	SGR	18 32.6	-25 08	12.2	3 s		2	15.9	Planetary, stellar
NGC 6818	PK 25-17.1	PLNNB	SGR	19 44.0	-14 09	10	22 s	15 s	4	15	B, vS, R
PK 1-6.2	SwSt 1	PLNNB	SGR	18 16.2	-30 52	11.8	<5 s		1		
PK 2-2.4	M2-23	PLNNB	SGR	18 01.7	-28 26	12.4	2 s		1		

!	remarkable object	!!	very remarkable object	E	elongated	
am	among	n	north	e	extremely	
att	attached	N	nucleus	er	easily resolved	
bet	between	neb	nebula, nebulosity	F	faint	
B	bright	P w	paired with	f	following	
b	brighter	p	pretty (before F, B, L, S)	g	gradually	
C	compressed	p	preceding	iF	irregular figure	
c	considerably	P	poor	inv	involved	
Cl	cluster	R	round	irr	irregular	
D	double	Ri	rich	L	large	
def	defined	r	not well resolved	l	little	
deg	degrees	rr	partially resolved	mag	magnitude	
diam	diameter	rrr	well resolved	M	middle	
dif	diffuse	S	small	m	much	

s	suddenly
s	south
sc	scattered
susp	suspected
st	star or stellar
v	very
var	variable
nf	north following
np	north preceding
sf	south following
sp	south preceding
11m	11th magnitude
8 . . . 13	8th mag and fainter
9 . . . 13	9th to 13th magnitude

Object	Other	Type	Con	R.A.	Dec.	Mag	SIZE_MAX	SIZE_MIN	Class	BRSTR	Description and notes
PK 2-5.1		PLNNB	SGR	18 14.6	-29 49	11.5	<10s				
PK 2-9.1	Canon 1-5	PLNNB	SGR	18 29.2	-31 30	11.9	<5s				
PK 3-4.7		PLNNB	SGR	18 11.6	-28 22	11	12s				
PK 3-4.9		PLNNB	SGR	18 12.8	-28 20	12	<10s		1		
PK 3-6.1	H2-43	PLNNB	SGR	18 17.7	-29 08	13	8.8s	5.2s	2		
PK 3-14.1	M2-36	PLNNB	SGR	18 55.6	-32 16	10.9	4s		2	18	
PK 3-17.1	Hubble 7	PLNNB	SGR	19 05.6	-33 12	12.9	2s		2	15.7	
PK 5-2.1	Hubble 8	PLNNB	SGR	18 07.9	-25 24	13	10s	10s	2		
PK 6+2.5		PLNNB	SGR	17 52.7	-22 22	13	25s		1		
PK 7+1.1	M1-31	PLNNB	SGR	17 55.1	-21 45	11	6.6s		2	14.7	
PK 11+4.1	Hubble 6	PLNNB	SGR	17 56.3	-16 30	12	8.0s	7.3s	2		
PK 12-7.1	M1-32	PLNNB	SGR	18 42.6	-21 17	12	<10s				
PK 12-9.1		PLNNB	SGR	18 50.5	-22 35	13	3.7s		2		
PK 13-3.1	M1-62	PLNNB	SGR	18 29.5	-19 06	13	4.9s	4.7s	2		
PK 16-4.1	M1-48	PLNNB	SGR	18 36.2	-17 00	13	17s	10s	3		
PK 355-6.5	M1-54	PLNNB	SGR	18 02.5	-36 39	11.7	<5s			13.5	
PK 358-5.1	M3-21	PLNNB	SGR	18 01.7	-33 15	13	9.9s	8.0s	2		
PK 358-6.1		PLNNB	SGR	18 09.9	-33 19	12	<10s				
PK 359-0.1	Hubble 5	PLNNB	SGR	17 47.9	-30 00	11.8	19s	12s	2?(6)		
B 7		DRKNB	TAU	04 33.0	+26 06	none	600m		6Ir		Narrow lanes E-W; Area contains B22-24 and B208-220
Be 84		DRKNB	TAU	04 22.1	+19 30	none	20m	10m	4Ir		Associated w nebula NGC 1554-5
IC 349	vdB 22	BRTNB	TAU	03 46.3	+23 56	none	30m		R		eF, vS, E 165 deg, dist 36" from Merope; small and faint inside Merope Neb
Mel 22	M 45	CL+NB	TAU	03 47.0	+24 07	1.2	100m		I 3 r n	2.9	vvB, vL, brilliant naked eye cluster, neb inv; Pleiades
NGC 1514	PK 165-15.1	PLNNB	TAU	04 09.3	+30 47	10.8	120s	90s	3(2)	9.5	*9 in neb 3'Diam

Name	Designation	Type	Con	RA	Dec	Mag	Size		Form		Remarks
NGC 1555	DG 31	BRTNB	TAU	04 21.9	+19 32	none	0.5 m		R		vF, S; Hind's Variable Nebula, associated with T Tauri
NGC 1952	M 1	SNREM	TAU	05 34.5	+22 01	8.4	8 m	4 m			vB, vL, E135, vglbM, r; Crab Nebula
Sh2-240	Simeis 147	BRTNB	TAU	05 39.1	+28 00	none	200 m	180 m	E		eL, eF; Huge oval cloud w filaments; probably supernova remnant
vdB 20		BRTNB	TAU	03 44.9	+24 07	11.6	20 m	16 m	R		Electra neby
vdB 23		BRTNB	TAU	03 47.5	+24 06	11.9	27 m	27 m	R		Alcyone neby
IC 4699	PK 348-13.1	PLNNB	TEL	18 18.5	-45 59	12	5 s		2	15.1	S, F, Stellar
NGC 5844	PK 317-5.1	PLNNB	TRA	15 10.7	-64 40	12	60 s				pB, pL, R, vgvlbM
NGC 5979	PK 322-5.1	PLNNB	TRA	15 47.7	-61 13	13	8 s			13	pF, vS, R, am 150st
PK 322-6.1	He2-136	PLNNB	TRA	15 52.3	-62 31	12.5	<10 s				
NGC 588		GX+DN	TRI	01 32.8	+30 39	none	0.7 m		Pec		F, p of 2; in M 33
NGC 592		GX+DN	TRI	01 33.2	+30 39	none	0.3 m		Pec		F, pL, f of 2; in M 33
NGC 595		GX+DN	TRI	01 33.6	+30 42	none	0.5 m		Pec		vF, S, R, inv in M 33
NGC 604		GX+DN	TRI	01 34.6	+30 47	none	2 m		Pec		B, vS, R, vvlBM; E Neb in NE arm of M 33
NGC 256		SMCCN	TUC	00 45.9	-73 30	12					F, S, R, gbM, * 9 nf 40"
NGC 299		SMCCN	TUC	00 53.4	-72 12	11.5					pB, vS, R, gvlbM, r

Abbr	Meaning	Abbr	Meaning	Abbr	Meaning	Abbr	Meaning
!	remarkable object	!!	very remarkable object	n	north	s	suddenly
am	among	E	elongated	N	nucleus	s	south
att	attached	e	extremely	neb	nebula, nebulosity	sc	scattered
bet	between	er	easily resolved	P w	paired with	susp	suspected
B	bright	F	faint	p	pretty (before F, B, L, S)	st	star or stellar
b	brighter	f	following	p	preceding	v	very
C	compressed	g	gradually	P	poor	var	variable
c	considercbly	iF	irregular figure	R	round	nf	north following
Cl	cluster	inv	involved	Ri	rich	np	north preceding
D	double	irr	irregular	r	not well resolved	sf	south following
def	defined	L	large	rr	partially resolved	sp	south preceding
deg	degrees	l	little	rrr	well resolved	11 m	11th magnitude
diam	diameter	mag	magnitude	S	small	8 . . .	8th mag and fainter
dif	diffuse	M	middle			9 . . . 13	9th to 13th magnitude
		m	much				

Object	Other	Type	Con	R.A.	Dec.	Mag	SIZE_MAX	SIZE_MIN	Class	BRSTR	Description and notes
NGC 306	ESO 29-SC23	SMCCN	TUC	00 54.3	-72 15	12.5					F, vS
NGC 346	ESO 51-SC10	SMCCN	TUC	00 59.1	-72 11	10.3	5.2 m				B, L, viF, mbMD*, r
NGC 3587	M 97	PLNNB	UMA	11 14.8	+55 01	11	202 s	196 s	3a	14	!! vB, vL, R, vvg, vsbM; Owl Nebula
NGC 5447		GX+DN	UMA	14 02.5	+54 17	none			Pec		pB, S, R, gmbM, conn M101
NGC 5449		GX+DN	UMA	14 02.5	+54 20	none			Pec		vF, pL, gvlbM, conn M101
NGC 5450		GX+DN	UMA	14 02.5	+54 16	none			Pec		F, pS, iR, glbM, conn M101
NGC 5451		GX+DN	UMA	14 02.6	+54 22	none			Pec		vF, pL, iR, vlbM, conn M101
NGC 5453		GX+DN	UMA	14 02.9	+54 18	none			Pec		F, pL, lE, vlbM, conn M101
NGC 5455		GX+DN	UMA	14 03.0	+54 14	none			Pec		pB, pS, R, psbM, conn M101
NGC 5461		GX+DN	UMA	14 03.7	+54 19	none			Pec		B, pS, R, psbM, conn M101
NGC 5462		GX+DN	UMA	14 03.9	+54 22	none			Pec		pB, pL, iR, gbM, conn M101
Gum 12		SNREM	VEL	08 30.0	-45 00	none	1200 m				Gum Nebula; pulsar PSR 0833-45 invol
NGC 2736	RCW 37	BRTNB	VEL	09 00.4	-45 54	none	30 m	7.0 m	E		! eeF, vL, vvmE 19 deg; filamentary nebula
NGC 2899	PK 277-3.1	PLNNB	VEL	09 27.1	-56 06	12.2	2.0 m			15.9	F, pL, R, gmbM, am 80st

				RA	Dec						
NGC 3132	PK 272+12.1	PLNNB	VEL	10 07.0	−40 26	8.2	84 s	53 s	4(2)	10.1	!! Planetary, vB, vL, lE *9m; Eight Burst Nebula
PK 261+2.1	He2-15	PLNNB	VEL	08 53.5	−40 04	13	20 s				
PK 264-8.1	He2-7	PLNNB	VEL	08 11.5	−48 43	12.4	25 s		2		
PK 265-2.1	Velghe 26	PLNNB	VEL	08 43.6	−46 06	13	<5 s		1		
PK 275-4.1	PB 4	PLNNB	VEL	09 14.9	−54 53	12.9	14 s	9 s			
PK 285+1.1	Peimbert 1-1	PLNNB	VEL	10 38.6	−56 47	8.6	<5 s				
PK 286-4.1	He2-55	PLNNB	VEL	10 48.8	−56 03	12.7	18 s				
PK 318+41.1	Abell 36	PLNNB	VIR	13 40.6	−19 53	13	478 s	281 s	3b(3a)	11.5	!! vB, vL, bi-N, lE, Dumbbell Neb
NGC 6853	M 27	PLNNB	VUL	19 59.6	+22 43	7.3	480 s	340 s	3(2)	14.1	
PK 72-17.1	Abell 74	PLNNB	VUL	21 16.8	+24 10	12.2	871 s	791 s	2	17.4	vF, L, irR, nBM at 100X; UHC filter helps
Sh2-88		BRTNB	VUL	19 46.0	+25 20	none	18 m	6 m	E		Has two small brighter knots w in large faint neby

!	remarkable object	n	north	s	suddenly
!!	very remarkable object	N	nucleus	s	south
am	among	neb	nebula, nebulosity	sc	scattered
att	attached	P w	paired with	susp	suspected
bet	between	p	pretty (before F, B, L, S)	st	star or stellar
B	bright	p	preceding	v	very
b	brighter	P	poor	var	variable
C	compressed	R	round	nf	north following
c	considerably	Ri	rich	np	north preceding
Cl	cluster	r	not well resolved	sf	south following
D	double	rr	partially resolved	sp	south preceding
def	defined	rrr	well resolved	11m	11th magnitude
deg	degrees	S	small	8 . . . m	8th mag and fainter
diam	diameter			9 . . . 13	9th to 13th magnitude
dif	diffuse				
E	elongated				
e	extremely				
er	easily resolved				
F	faint				
f	following				
g	gradually				
iF	irregular figure				
inv	involved				
irr	irregular				
L	large				
l	little				
mag	magnitude				
M	middle				
m	much				

Index

Index